BRITISH GEOLOGICAL SURVEY

A H COOPER and
I C BURGESS

# Geology of the country around Harrogate

Memoir for 1:50 000 geological sheet 62
(England and Wales)

CONTRIBUTORS

*Stratigraphy*
A C Benfield
H Johnson

*Hydrocarbons*
D W Holliday

*Hydrogeology*
R A Monkhouse

*Mineral springs*
W M Edmunds

*Palaeontology*
D K Graham
H C Ivimey-Cook
J Pattison
W H C Ramsbottom
N J Riley
A R E Strank
G Warrington
I P Wilkinson

LONDON: HMSO    1993

First published 1993

ISBN 011 884489 X

Bibliographical reference

COOPER, A H, and BURGESS, I C. 1993.   Geology of the country around Harrogate.   *Memoir of the British Geological Survey*, sheet 62 (England and Wales).

*Authors*

A H Cooper, BSc, PhD
*British Geological Survey, Newcastle*

I C Burgess, BSc
*Formerly British Geological Survey*

*Contributors*

A C Benfield, BSc
W H C Ramsbottom, MA, PhD
A R E Strank, BSc, PhD
*Formerly British Geological Survey*

D W Holliday, MA, PhD,
H C Ivimey-Cook, BSc, PhD
J Pattison, MSc
N J Riley, BSc, PhD
G Warrington, BSc, PhD
I P Wilkinson, MSc, PhD
*British Geological Survey, Keyworth*

D K Graham, BA
H Johnson, BSc
*British Geological Survey, Edinburgh*

W M Edmunds, BSc, PhD
R A Monkhouse, BA, MSc
*British Geological Survey, Wallingford*

Printed in the UK for HMSO
Dd 292032  C8  03/93

*Other publications of the Survey dealing with this and adjoining districts*

BOOKS

*British Regional Geology*
The Pennines and adjacent areas, 3rd edition, 1954
Eastern England, 2nd edition, 1980

*Memoirs*
Geology of the country between Bradford and Skipton (Sheet 69), 1953
Geology of the country north and east of Leeds (Sheet 70), 1950
Geology of the country around Thirsk (Sheet 52), 1992

*Mineral Assessment Reports*
The sand and gravel resources of the country east of Harrogate (Sheet SE 35)    Report 70, 1981
The sand and gravel resources of the country west of Boroughbridge (Sheet SE 36)    Report 78, 1981
The sand and gravel resources of the country around Kirk Hammerton (Sheet SE 45)    Report 84, 1981
The sand and gravel resources of the country around Tholthorpe (Sheet SE 46)    Report 88, 1982
The sand and gravel resources of the country around West Tanfield (Sheet SE 27)    Report 135, 1983
The sand and gravel resources of the country north-east of Ripon (Sheet SE 37 and part of SE 47)    Report 143, 1984

*Open File Reports*
The Permian Rocks of the Harrogate district. Geological notes and local details of 1:50 000 Sheet 62, 1987
The Permian rocks of the Thirsk district. Geological notes and local details of 1:50 000 sheet 52, 1987
The geology of the country north and east of Ripon, North Yorkshire, with particular reference to the sand and gravel deposits; description  of 1:25 000 sheet SE 37, 1983
The geology of the country around Dalton, North Yorkshire, with particular reference to sand and gravel deposits: description of 1:25 000 sheet SE 47, 1983
The sand and gravel resource of the Wharfe Valley between Ilkley and Collingham, 1984

MAPS

*1:625 000*
Solid Geology (south sheet), 1979
Quaternary Geology (south sheet), 1977

*1:250 000*
(*Solid Geology, Quaternary, Aeromagnetic and Gravity Sheets*)
Tyne Tees
Humber-Trent (no Quaternary map)

*1:100 000*
Hydrogeological map of southern Yorkshire and adjoining areas, 1982

*1:50 000 (and one inch to one mile)*
Sheet 51    Masham (Solid), 1985
Sheet 51    Masham (Drift), 1985
Sheet 52    Thirsk (new edition in preparation)
Sheet 53    Pickering (Solid and Drift), 1973
Sheet 61    Pateley Bridge (out of print)
Sheet 62    Harrogate (Solid), 1987
Sheet 62    Harrogate (Drift), 1987
Sheet 63    York (Solid and Drift), 1983
Sheet 69    Bradford (Solid), 1974
Sheet 69    Bradford (Drift), 1974
Sheet 70    Leeds (Solid), 1974
Sheet 70    Leeds (Drift), 1974
Sheet 71    Selby (Solid and Drift), 1973

# CONTENTS

TABLES

PLATES

# PREFACE

The resurvey of the Harrogate district was intended primarily to assess its economic mineral wealth, notably sand and gravel; it was carried out between 1973 and 1978 with funds provided largely by the Department of the Environment. The resurvey also yielded important information on the Permian limestone resources, which are a valuable source of aggregate for the construction industry. At around the same time, the National Coal Board started an investigation of the North Ouse Coalfield within the district and this has yielded much new information on the Carboniferous and Permian sequences, and on the structure of the district.

Not only has this resurvey contributed to our knowledge of the resources of the area, but it has also produced important new scientific information. For example, the Harrogate area straddles the margin between the Askrigg Block and the Craven Basin; the marked thickness changes in the Carboniferous sequence across this line, similar to those already described from areas farther west, are recognised for the first time in the district. The localisation of tectonic structures along this line in the Harrogate district is also newly recognised; the eastward continuation of it marks the northern margin of the North Ouse Coalfield, currently under development. These structures were later reactivated and also affect the Jurassic and Permian rocks of the area. The classic unconformity between the Carboniferous and Permian sequences is described and illustrated, as are newly recognised thickness and facies distributions for the Permian rocks. Previously unknown belts of desert sand dunes buried at the base of the Permian sequence are described from the east of the district. These were covered by cyclic deposits of dolomite, anhydrite/gypsum and salt, laid down in the evaporitic Zechstein sea. The overlying terrestrial redbeds of the Triassic were later inundated and buried beneath marine deposits in the Jurassic.

For the last fifty years almost nothing has been written about the glacial deposits of the district. This has been redressed with this modern account of the Quaternary glaciation of the area. Previously unknown facies associations in buried valleys are described and the extent of the buried sand and gravel deposits explained. The complex nature of the glacial till, the relationships between the esker and moraine deposits, and the patterns of marginal channels and outwash deposits are related to the stagnation and retreat of the Devensian ice sheet. The development of extensive glacial lake deposits is also documented.

Potential sources of geological hazard are identified, including a natural subsidence problem in the Ripon area caused by the dissolution of gypsum. These factors are important for planning and development in the district.

This memoir shows how a field survey study, initiated to assess sand and gravel resources, can also yield other commercial and scientific results of relevance to mineral extraction companies, planners, developers and academics.

Peter J Cook, DSc
*Director*

*British Geological Survey*
*Kingsley Dunham Centre*
*Keyworth*
*Nottingham*
*NG12 5GG*

November 1992

# HISTORY OF SURVEY OF THE HARROGATE SHEET

# NOTES

The district was originally surveyed on the scale of six inches to one mile by W T Aveline, J R Dakyns and C Fox-Strangways between 1869 and 1871. It was published on the one-inch scale as Old Series sheet 93NW (Solid and Drift) in 1874. The related memoir by C Fox-Strangways, 'The geology of the country north and east of Harrogate', was published in 1874 with a second edition in 1908.

In 1973–78 the district was resurveyed on the six-inch or 1:10 000 scale by Drs A H Cooper, G D Gaunt and R Anderton and by Messrs I C Burgess, H Johnson, A C Benfield, J P B Szymanski, D H Land and J S C Sceal. The six-inch or 1:10 000 maps covered by the various surveyors are listed with the acknowledgements. The drift and solid editions of the 1:50 000 map of the district were published in 1987.

Throughout the memoir the word 'district' refers to the area covered by the 1:50 000 Harrogate (62) Sheet.

National Grid references are given in square brackets unless otherwise stated; all lie within the 100 km square SE.

Bed thickness descriptions are based on the classification of Ingram (1954).

# ACKNOWLEDGEMENTS

This memoir has largely been written by Dr A H Cooper with help from Mr I C Burgess. Mr A C Benfield contributed stratigraphical information on the Triassic, Jurassic and glacial deposits. Mr H Johnson provided details of some Carboniferous and Permian sections, along with much information about the glacial deposits. Dr W M Edmunds provided information on the Harrogate springs and Mr R A Monkhouse the hydrogeological notes. Dr D W Holliday contributed the section on hydrocarbon potential. The Carboniferous faunas were mainly identified by Dr W H C Ramsbottom, who named most of the goniatites and provided the biostratigraphical correlations, and Mr J Pattison who identified some of the goniatites and most of the benthonic faunas. Some Carboniferous faunas were also identified by Drs N J Riley and A R E Strank. Mr J Pattison identified the Permian faunas, Dr G Warrington the late Triassic palynomorph assemblages, Dr I P Wilkinson the Jurassic foraminifera and Dr H C Ivimey-Cook the Jurassic macrofossils.

Much new data about the Quaternary deposits, rockhead and concealed geology were accrued from the Survey's mineral assessment surveys; the contributions made by Messrs D A Abraham, D L Dundas, J R A Giles, J W C James, A N Morigi and R Stanczyszyn in this field are acknowledged. Comments from Mr M Culshaw about the engineering geology aspects of the memoir are also acknowledged. The memoir was edited by Messrs J I Chisholm and R J Wyatt.

Geological National Grid six-inch and 1:10 000 scale maps included wholly or in part in Sheet 62 are listed below with the initials of the surveyors and the dates of the survey. The surveyors were R Anderton, A C Benfield, I C Burgess, A H Cooper, G D Gaunt, H Johnson, D H Land, J S C Sceal and J P B Szymanski for maps contained wholly within Sheet 62, and C G Godwin, J H Powell, J G O Smart and A A Wilson for maps which include peripheral areas of the sheet. All the maps are available as dyeline prints from BGS Keyworth.

| | | | |
|---|---|---|---|
| SE 25 NE | Killinghall | ICB | 1977–1978 |
| SE 25 SE | Pannal Ash | ICB | 1977–1978 |
| SE 26 NE | Fountains Abbey | HJ | 1976–1978 |
| SE 26 SE | Bishop Thornton | AHC | 1977–1978 |
| SE 27 SE | Galphay | AHC, HJ, JGOS, AAW | 1975–1980 |
| SE 35 NW | Harrogate | ICB | 1977–1978 |
| SE 35 NE | Knaresborough | AHC | 1976 |
| SE 35 SW | Follifoot | ICB | 1977–1978 |
| SE 35 SE | Spofforth | AHC | 1977–1978 |
| SE 36 NW | Bishop Monkton | RA, AHC | 1973–1977 |
| SE 36 NE | Boroughbridge | RA | 1974 |
| SE 36 SW | Burton Leonard | JSCS, AHC | 1974–1977 |
| SE 36 SE | Staveley | JSCS, AHC | 1973–1977 |
| SE 37 SW | Ripon | AHC | 1980–1981 |
| SE 37 SE | Dishforth | RA, CGG | 1974–1980 |
| SE 45 NW | Whixley | GDG | 1975 |
| SE 45 NE | Green Hammerton | RA | 1975 |
| SE 45 SW | Hunsingore | AHC, GDG | 1976 |
| SE 45 SE | Tockwith | AHC | 1975–1976 |
| SE 46 NW | Aldborough | JPBS | 1976 |
| SE 46 NE | Tholthorpe | HJ | 1976 |
| SE 46 SW | Marton Cum Grafton | JPBS | 1975–1976 |
| SE 46 SE | Aldwark | HJ | 1975–1976 |
| SE 47 SW | Cundall | HJ, ACB | 1978–1982 |
| SE 47 SE | Raskelf | HJ, ACB | 1977–1979 |
| SE 55 NW | Beningbrough | RA, JPBS | 1975–1977 |
| SE 55 NE | Skelton | HJ, DHL | 1976–1978 |
| SE 55 SW | Rufforth | DHL | 1976 |
| SE 55 SE | Nether Poppleton | HJ, DHL | 1976–1979 |
| SE 56 NW | Easingwold | ACB | 1975–1977 |
| SE 56 NE | Stillington | HJ | 1977–1978 |
| SE 56 SW | Tollerton | HJ, JPBS | 1977 |
| SE 56 SE | Sutton on the Forest | HJ | 1977 |
| SE 57 SW | Oulston | ACB, JHP | 1981–1984 |
| SE 57 SE | Crayke | HJ, ACB, JHP | 1978–1984 |

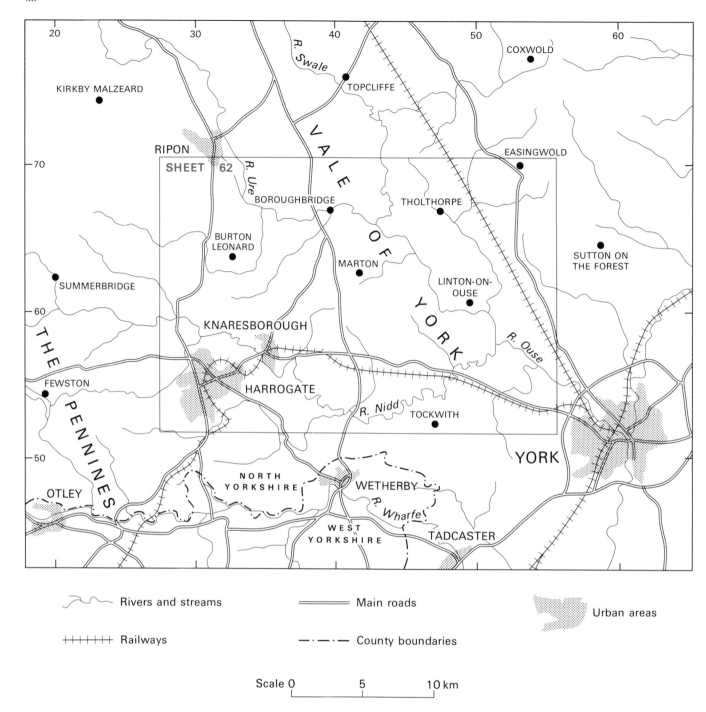

**Figure 1**  Physical features and location of the Harrogate district.

# ONE

# Introduction

## GEOGRAPHICAL SETTING

The district described in this memoir lies in North Yorkshire and is covered by Sheet 62 of the 1:50 000 Geological Map Series of England and Wales. The spa town of Harrogate and the market towns of Knaresborough, Boroughbridge and Ripon are the main population centres; the remainder of the district is rural. Most of the land is given over to arable agriculture, with the high ground in the south-west mainly pastoral. Industry is restricted to a few small sites around Harrogate, which is a popular holiday and conference centre founded on the Victorian hotels that developed around the mineral springs. Commercial exploitation of other geological resources is limited to magnesian limestone quarrying near Markington, and sand and gravel extraction around Knaresborough and Staveley.

The principal physiographical feature of the district (Figure 1) is the Vale of York, which trends north-north-west between the Pennines, close by to the west, and the North York Moors and Yorkshire Wolds, away to the east. The highest ground is in the south-west, rising to around 170 m above OD, coincident with the resistant Carboniferous strata of the Harrogate Anticline. The dominant north-north-westerly grain of the landscape reflects the gentle easterly dip of the late Palaeozoic and early Mesozoic strata (Figure 2). The Permian limestones form two marked north-north-west-trending escarpments flanking the Carboniferous rocks of the Pennines. Farther east the soft Permian and Triassic mudstones and sandstones have been eroded to form the Vale of York, where they are heavily blanketed with glacial deposits. In the north-east corner of the district, relatively resistant Jurassic strata form the rising ground leading up to the North York Moors.

## GEOLOGICAL SEQUENCE

The geological formations that crop out in the district are listed on the inside of the front cover. Information on the older rocks is sparse, and is based on a single borehole and inference from geophysical explorations.

## OUTLINE OF GEOLOGICAL HISTORY

Little is known about the pre-Carboniferous basement rocks of the Harrogate district, which lies east of the Ordovician-Silurian Craven inliers (Arthurton et al., 1988) and north-west of sequences of similar age, which are peripheral to the East Midlands microcraton (Pharaoh et al., 1987). However, on the grounds of regional analogy, it is thought likely that the basement comprises Ordovi-cian and Silurian strata folded by the Acadian deformation (formerly called Caledonian in this area) which imparted the basement grain to the whole of northern England (Soper et al., 1987). Granites intruded during the same period (Dunham, 1974) produced relatively buoyant areas such as the Askrigg Block, and differentiated these from intervening less buoyant basinal areas. The Harrogate district was situated in a position which, in Carboniferous times, spanned the transition from the stable Askrigg Block in the north-west to the subsiding Craven and Cleveland basins to the west and north-east respectively (Figure 4). This basement control influenced the Carboniferous depositional history and subsequent Hercynian deformation.

In the district, information on Dinantian rocks is limited to the Ellenthorpe Borehole, and sections in the core of the Harrogate Anticline. These mainly prove the existence of a basinal sequence of limestones and mudstones, similar to those of the Craven Basin. The overlying Namurian rocks, although of limited exposure, show a marked southward and eastward thickening away from the flanks of the Askrigg Block. The Namurian sequence is roughly cyclic with numerous northerly derived fluviatile to deltaic sandstones interbedded with marine mudstones. These alternations record the influence of varying sea level, which resulted from an interaction between tectonic effects and world-wide eustatic sea-level fluctuations (Ramsbottom, 1966; Leeder, 1988). Deltaic and fluvial conditions persisted, although interrupted by numerous marine transgressions, through Namurian and into Westphalian times. The Westphalian strata show a variable cyclicity between marine, deltaic, swamp and subaerial environments (Guion and Fielding, 1988; Rowley et al., 1985) in a near equatorial climate. Numerous coals developed from the swamp peats which formed on fluviodeltaic plains near to sea level.

Throughout the Carboniferous, the district was probably subject to sporadic earth movements of the Hercynian orogeny, with a climax involving inversion of the depositional basins in the late Carboniferous (Asturian phase). Compression and dextral strike slip or transpression (Harland, 1971) affected the region, the effects being most evident along the block margins. A major fold belt, initiated in the Dinantian, developed along the southern edge of the Askrigg Block, extending from Settle (Arthurton, 1984; Arthurton et al., 1988) eastwards to the Harrogate district. Here, the Harrogate Anticline developed, along with numerous minor folds in the adjacent ground to the north.

The start of the Permian period saw a terrestrial palaeogeography and a desert environment, with subaerial erosion of the newly uplifted Hercynian structures. Highs such as the Harrogate Anticline were eroded and the debris transported by fluvial and aeolian processes

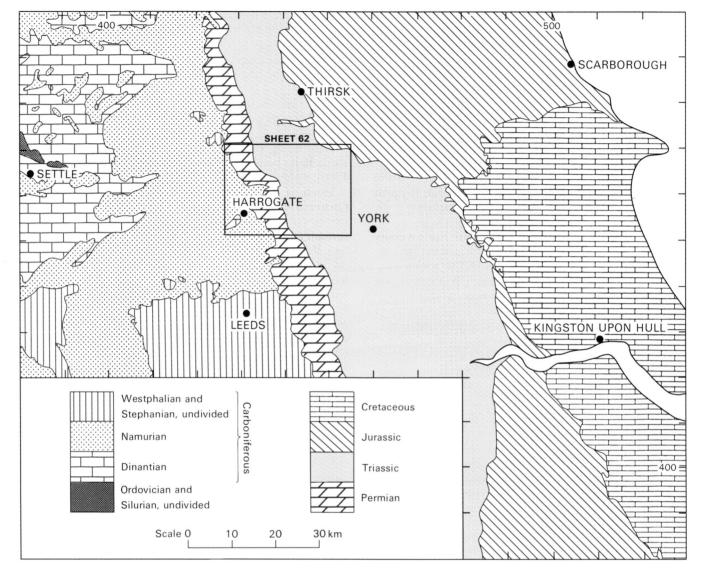

**Figure 2**  Generalised solid geology of the Harrogate district.

into a major land-locked basin which extended east-wards. Resistant sandstone areas were sculpted into steep-sided hills; the softer siltstones and mudstones weathered down to form undulating areas of stony desert with thin breccia deposits. Wind-blown sand dunes migrated across this landscape until late Permian times.

In the late Permian, rifting along the central North Sea line caused a seaway to open to the north, allowing the enclosed basin to be rapidly flooded by the Zechstein sea (Smith, 1980, 1990). Initial sedimentation was euxinic and resulted in the deposition of widespread thin carbonaceous and calcareous siltstones, the Marl Slate (represented here by the Lower Marl). This sedimentary unit formed throughout most of the basin. As marine conditions stabilised, the district became the locus of a carbonate platform flanked by reefs to the east (the Lower Magnesian Limestone). This was followed by less-marine conditions, possibly related to restricted availability of sea water through the northern

supply channel. The Zechstein sea became progressively more saline as the water supply decreased and evaporation concentrated the dissolved salts; anhydrite and gypsum were deposited mainly in marginal sabkhas and hypersaline lagoons, followed by halite and sylvinite, and then by subaerial redbed deposition. This first cycle of flooding and evaporation was followed by four more, forming the English Zechstein Cycles EZ1–EZ5. Not all the cycles were as complete and widespread as the first, but limestones developed in three of them. Only the limestones of cycles EZ1 and EZ3 (the Lower and Upper Magnesian Limestones; also called the Cadeby and Brotherton formations) are well exposed in the district. The highly soluble evaporite minerals in the Permian sequence were concentrated mainly in the more central parts of the basin. Their lateral equivalents near outcrop are the red-brown mudstones, of the Middle Marl (Edlington Formation) and Upper Marl (Roxby Formation).

Towards the end of the Permian, the Zechstein sea finally dried up. The early Triassic saw a landscape of low relief and an abrupt change to semidesert conditions. In this environment fluviatile redbed deposition (Sherwood Sandstone Group; formerly Bunter Sandstone), was succeeded by playa lake or desert plain deposits of red-brown dolomitic mudstones, with subordinate gypsum (Mercia Mudstone Group; formerly Keuper Marl). In the late Triassic this facies was followed by the deposition of grey-green dolomitic mudstones (Blue Anchor Formation) marking a change to a coastal lagoon (salina) environment, a precursor to the marine transgression of the overlying Penarth Group. The lower part of the Penarth Group (Westbury Formation) comprises carbonaceous, grey-black shales and bioturbated sandstones. These organic-rich sediments have an abundant low-diversity fauna and represent deposition in a shallow sea with an anaerobic substrate. The upper part of the group (Lilstock Formation; here represented solely by the Cotham Member) is a thin sequence of calcareous mudstones deposited in brackish water. It represents a shallowing of the environment before the onset of fully marine conditions in the overlying Lias Group.

The Triassic–Jurassic boundary is recognised near the base of the Redcar Mudstone Formation, the lowest unit of the Lias Group. This formation of deep-water mudstones and limestones passes upwards into the shallow-water sandstones of the Staithes Formation. Continued shallow-water conditions produced the oolitic-'chamositic' ironstones of the Cleveland Ironstone Formation. The overlying Whitby Mudstone Formation, which is the youngest Jurassic unit seen in the district, marks a return to fully marine deep-water mudstones.

Much of the district was probably covered by a full succession of Jurassic and Cretaceous rocks similar to those seen in the North York Moors and Yorkshire Wolds. It was subject to mid-Cretaceous earth movements, which reactivated basement structures and caused faulting, especially along the continuation of the Howardian Hills Fault Belt, between Harrogate and Easingwold (Kent, 1980); uplift and erosion followed during the Tertiary.

Much of the district is mantled by Drift deposits of Quaternary age. The bulk of these are of glacial and periglacial origin and may be attributed to the Late Devensian cold stage (approximately 18 000 to 14 000 years ago; Rose, 1985; Bowen et al., 1986). Some older glacial till of unknown affinity may also be present in the southwest of the district, and pre-Devensian gravels and clays partly fill some of the larger buried valleys. Devensian till blankets most of the district and is associated with numerous esker and moraine ridges. Supraglacial and periglacial lake deposits abound, in complex facies associations that reflect the fluctuating advance and decay of the icesheet. Postglacial (Flandrian) terrace and alluvial deposits locally overprint the Devensian features. Recently active geological processes include landslipping and subsidence due to groundwater dissolution of gypsum, mainly in the Permian sequence.

TWO

# Carboniferous

Carboniferous rocks crop out over an area of about 100 km² in the west of the district (Figure 3) and are present elsewhere at depth. Millstone Grit occupies most of the outcrop, with a small area of Carboniferous Limestone on the axis of the Harrogate Anticline and a faulted patch of Westphalian north of Knaresborough. In the Vale of York, where the Carboniferous rocks are concealed beneath Permian, Triassic and Jurassic rocks, boreholes prove that Coal Measures strata are present in the south-east and that Carboniferous Limestone occurs at Ellenthorpe; Millstone Grit is inferred elsewhere (Figure 6).

Early publications referring to Carboniferous rocks are concerned mainly with the Harrogate mineral springs; geological comments are incidental (see bibliography in Fox-Strangways, 1908). The first purely geological description was that of John Phillips (1865), who outlined the main features of the Harrogate Anticline, with accompanying geological sections. The main elements of the geology were delineated by the mapping of the Primary Geological Survey (Fox-Strangways, 1874, 1885, 1908). The sequence was further described by Hudson (1930, 1934, 1938a, 1944), who added palaeontological detail to the mainly lithological classifications of previous studies. However, the classifications proposed by Hudson were based on a misunderstanding of the sequence of strata exposed on the Harrogate Anticline; the names were used in differing senses, both in different parts of the ground and in his various publications. These early works formed the basis for the present resurvey.

The stratigraphical nomenclature used on the published 1:50 000 geological map is an informal system in which sandstones, coal seams, limestones and some fossil bands are named, but the intervening siltstones and mudstones are not. However, a more comprehensive lithostratigraphical nomenclature exists for adjacent areas to the north-west (summary in Dunham and Wilson, 1985) and reference to these names is included in the text where appropriate. On the map the basic colour scheme reflects the chronostratigraphical divisions; in ascending sequence, the Carboniferous Limestone is assigned to the late Dinantian (Viséan), the Millstone Grit to the Namurian, and those of the Coal Measures to the Westphalian.

Harrogate lies to the east of the classic region between Pateley Bridge and Settle, where the control of Carboniferous sedimentation by basement structures was first recognised by Marr (1921). Although many of the original concepts have since been modified (Kent, 1966; Ramsbottom, 1977; Collinson, 1988; Lee, 1988; Ebdon et al., 1990), particularly with relation to the concept of plate tectonics (Leeder, 1982b, 1988), the distinction between the Askrigg 'Block' to the north and the Craven 'Basin' to the south remains fundamentally valid (Figures 4 and 5). The Askrigg Block was an area of slow subsidence characterised by shallow-water sedimentation; the Craven Basin was an area of more rapid subsidence characterised at times by relatively deep-water sedimentation. This structural control of the deposition was most marked in the Dinantian (Leeder, 1988); it diminished upwards through the Namurian (Figure 7). As Dinantian and early Namurian rocks have a restricted outcrop in the Harrogate district, the precise facies boundary between block and basin is difficult to identify here and, indeed, it may have migrated with time. Evidence from commercial seismic reflection data suggests that the facies boundary may be related to a number of concealed east-north-east-trending syndepositional faults in the north-west corner of the district, to the north of the Ripley–Staveley Fault Belt (Figure 6). These dislocations are probably the easterly continuations of the Mid Craven Fault of the Settle district (Arthurton et al., 1988).

The information on pre-Brigantian rocks is insufficient to support any conclusions concerning their palaeogeography. However, during Brigantian and early Pendleian times a deep basin, in which turbidites and goniatite-bearing mudstones were laid down, formed an eastward extension of the Craven Basin in the area around and to the south of Harrogate (Figure 4). This basin was filled during the late Pendleian by an influx of clastic detritus; beds of this age were subsequently uplifted and eroded during a short period of tectonism (Arthurton, 1983; Arthurton et al., 1988). They are overlain by the fluviatile Almscliff Grit (equivalent to the Brennand or Grassington grits), also of Pendleian age. The extent of this erosional episode, which is best documented on the Askrigg Block (Dunham and Wilson, 1985; Arthurton et al., 1988), diminishes into the Craven Basin (Ramsbottom, 1966; Ramsbottom et al., 1978). Evidence from around Pateley Bridge (Dunham and Wilson, 1985) and Masham (Wilson, 1960) indicates that, on the southern margin of the Askrigg Block, late Brigantian to early Pendleian limestones were eroded and unconformably overlain by the Grassington Grit (Figure 7, column 1). The same relationship apparently exists also in the north-west of the Harrogate district (Figure 7, column 2). Here, a thin Pendleian to Kinderscoutian sequence, known partly from the Aldfield and Sawley boreholes just within the Pateley Bridge district (Falcon and Kent, 1960), rests (unconformably?) on limestones of Brigantian age at Sawley. Southwards, the Pendleian and Arnsbergian sequence thickens dramatically and is more than four times as thick in the south of the district (Figure 7, column 6). The sediments of Chokierian and Kinderscoutian age thicken slightly southwards; the intervening Alportian strata are apparently missing (Ramsbottom, 1977). The late Namurian (Marsdenian and Yeadonian) sequence more than doubles its thickness

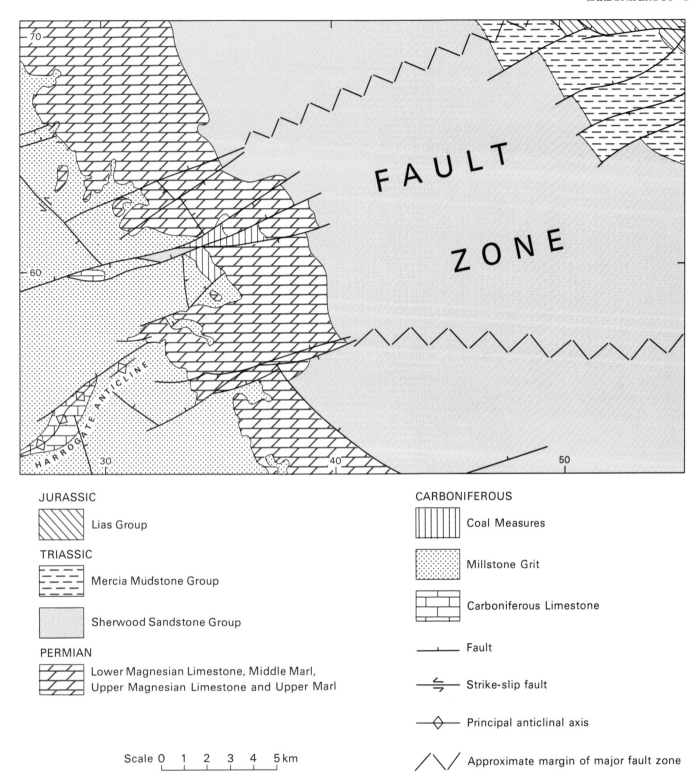

**Figure 3**  Geological sketch map of the Harrogate district.

southwards across the area (Figure 9). Most of the thickness change occurs across the strongly folded ground around Ripley, between the Cayton Gill Anticline and the Farnham Syncline (Figure 6). To the north of this belt the structure of the Carboniferous rocks is gently undulating, and south of the Harrogate Anticline it is also simple. The coincidence of this fold belt with the thickness changes places the southern margin of the Askrigg Block at the north side of the Cayton Gill Anticline (Figures 4 and 6).

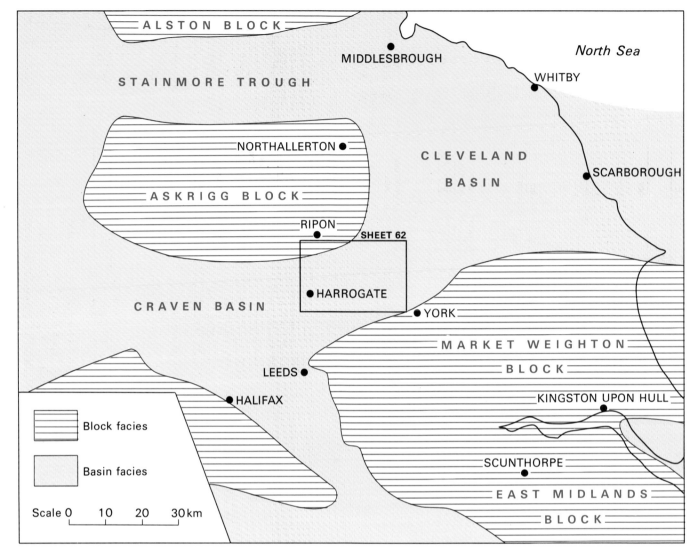

**Figure 4** Palaeogeographical setting of the Harrogate district with respect to early Carboniferous (mainly Viséan) blocks and basins (after Kent, 1966, 1974, 1980; Bott, 1978; Ebdon et al., 1990).

Overprinting the block–basin framework outlined above, the Namurian, from the late Pendleian Almscliff Grit onwards, is characterised by cyclothemic sedimentation (Holdsworth and Collinson, 1988). Each cyclothem was initiated by a rise in sea level, accompanied by an influx of marine faunas, and when fully developed comprises:

7  Coal
6  Seatearth
5  Sandstone, fine- to coarse-grained
4  Siltstone with sandstone beds and laminae
3  Mudstone and siltstone with ironstone nodules
2  Mudstone with marine fossils
1  Shelly mudstone, shelly bioturbated sandstone or muddy limestone.

Some marine bands contain dominantly shelly benthonic fossils, others contain mainly nektopelagic faunas, and some contain *Lingula* alone, reflecting variations in the depth, salinity and oxygen potential of the depositional environment. The dominant sediments are dark grey mudstone and siltstone, and some cycles consist only of units 1 to 3. The higher units in each cycle mark the encroachment of deltaic conditions, which introduced shoreface and delta front sands, and, as the sedimentary interface was built up above sea level, fluviatile sands, locally capped by soils and forests. Within this cyclic succession there is a major hiatus, representing the Alportian Stage, between the Upper Follifoot Grit and the overlying fossiliferous mudstones (Ramsbottom, 1974).

The early Westphalian rocks, known from the Farnham Borehole, are finer grained than the Namurian strata. The later Westphalian rocks, in the concealed coalfield in the York area (Moss, 1985), are a northward continuation of the well-known succession of the Yorkshire

| Stage | HARROGATE | LEEDS | BRADFORD-SKIPTON | SETTLE | MASHAM |
|---|---|---|---|---|---|
| WESTPHALIAN A | | | | | NOOK HOUSE SANDSTONE / NOOK HOUSE SHALE / WINKSLEY SANDSTONE / WINKSLEY SHALE |
| YEADONIAN | ROUGH ROCK / ROUGH ROCK FLAGS | ROUGH ROCK / ROUGH ROCK FLAGS | ROUGH ROCK / ROUGH ROCK FLAGS | ROUGH ROCK | LAVERTON SANDSTONE / LAVERTON SHALE |
| MARSDENIAN | HUDDERSFIELD WHITE ROCK / GUISELEY GRIT / BRANDON GRIT / EAST CARLTON GRIT | WHITE ROCK / GUISELEY GRIT / BRANDON GRIT / EAST CARLTON GRIT | HUDDERSFIELD WHITE ROCK / GUISELEY GRIT / WOODHOUSE/BRANDON GRIT / BLUESTONE / EAST CARLTON GRIT | GRETA GRIT | WANDLEY GILL SANDSTONE / WANDLEY GILL SHALE |
| KINDERSCOUTIAN | UPPER PLOMPTON GRIT / BILTON COAL / LOWER PLOMPTON GRIT / ADDLETHORPE GRIT / CAYTON GILL SHELL BED | BRAMHOPE/UPPER PLUMPTON GRIT / ECCUP MARINE BAND / CALEY CRAGS*/LOWER PLUMPTON GRIT / ADDLETHORPE GRIT / CAYTON GILL BEDS | BRAMHOPE GRIT / ADDINGHAM EDGE GRIT / CALEY CRAGS GRIT* / OTLEY SHELL BED / ADDLETHORPE GRIT* | ELDROTH GRIT / KNOTT COPY GRIT / ACCERHILL SANDSTONE | UPPER BRIMHAM GRIT / BRIMHAM SHALE / LOWER BRIMHAM GRIT / BEVERLEY SHALES / LIBISHAW SHELL BED / LIBISHAW SANDSTONE / LIBISHAW SHALE / AGILL SANDSTONE ON AGILL SHALE / CAYTON GILL SHELL BED / CAYTON GILL SHALE |
| ALPORTIAN | non sequence | non sequence | non sequence | non sequence | non sequence |
| CHOKIERIAN | UPPER FOLLIFOOT GRIT | UPPER FOLLIFOOT GRIT | BROCKA BANK GRIT | | UPPER FOLLIFOOT GRIT / FOLLIFOOT SHALE |
| ARNSBERGIAN | LOWER FOLLIFOOT GRIT / COLSTERDALE MARINE BEDS / RED SCAR GRIT | LOWER FOLLIFOOT GRIT | MIDDLETON GRIT / NESFIELD SANDSTONE / MARCHUP GRIT | SILVER HILLS SANDSTONE / KEASDEN FLAGS / CATON SHALES | LOWER FOLLIFOOT GRIT / SCAR HOUSE BEDS / COLSTERDALE MARINE BEDS / RED SCAR GRIT / NIDDERDALE SHALES |
| PENDLEIAN | ALMSCLIFF GRIT / HARROGATE ROADSTONE | ALMSCLIFF GRIT / LINDLEY MOOR GRIT / UPPER BOWLAND SHALES | SKIPTON MOOR GRITS / UPPER BOWLAND SHALES / BERWICK LIMESTONE | ROEBURNDALE FORMATION / BRENNAND/GRASSINGTON GRIT / non sequence / PENDLE GRIT FORMATION / UPPER BOWLAND SHALES | GRASSINGTON GRIT / non sequence / CROW LIMESTONE / RICHMOND CHERTS / LITTLE LIMESTONE / MAIN LIMESTONE |
| BRIGANTIAN | HARLOW HILL SANDSTONE | | MIDDLE BOWLAND SHALES / NETTLEBER SANDSTONE / LOWER BOWLAND SHALES | LOWER BOWLAND SHALES WITH PENDLESIDE SANDSTONE / non sequence | UNDERSET, THREE YARD, FIVE YARD, MIDDLE, SIMONSTONE, HARDROW SCAR, GAYLE AND HAWES LIMESTONES |
| ASBIAN | Un-named limestone | | DRAUGHTON SHALES / DRAUGHTON LIMESTONE / non sequence | PENDLESIDE LIMESTONE / non sequence | DANNY BRIDGE LIMESTONE/ KINGSDALE LIMESTONE |
| HOLKERIAN | | | SKIBEDEN SHALES | WORSTON SHALES | HORTON LIMESTONE |
| ARUNDIAN | Un-named limestone | | EMBSAY LIMESTONE (part) | | |

NAMURIAN · VISÉAN

ELLENTHORPE BOREHOLE ?

† BLOCK AND BASIN APPLY MAINLY TO VISÉAN

ASKRIGG BLOCK† · C R A V E N   B A S I N†

LOCATION OF SECTIONS

| | Masham | |
| Settle | Pateley Bridge no information | Harrogate |
| | Bradford-Skipton | Leeds |

**Figure 5** Chronostratigraphical correlation diagram for the Viséan and Namurian showing the approximate positions of the main named lithological units in the Harrogate district and surrounding areas; * denotes incorrectly named and correlated sandstone sequences. Based on Edwards et al. (1950); Stephens et al. (1953); Arthurton et al. (1988); British Geological Survey (1985); Burgess and Cooper, 1980b; George et al. (1976); Ramsbottom (1977) and Ramsbottom et al. (1978).

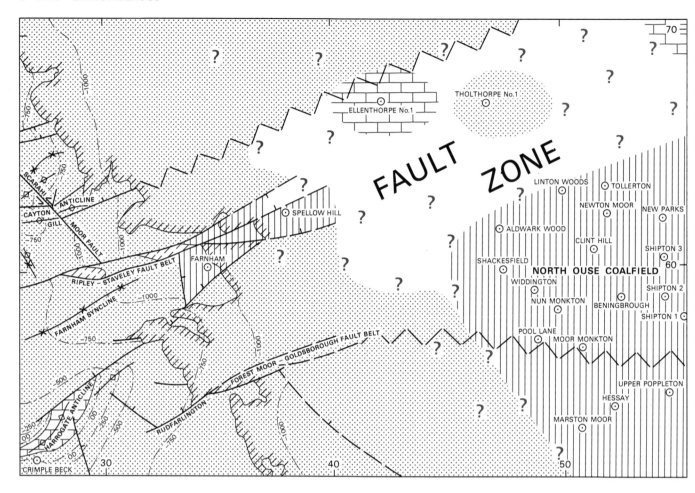

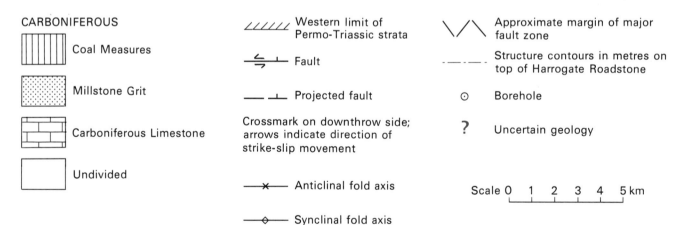

**Figure 6** Distribution of Carboniferous rocks at outcrop and at the sub-Permian unconformity. North Ouse Coalfield information from Moss (1985) and Whittaker (1985). Major named structures in the Carboniferous rocks are also shown, along with postulated structure contours on the top of the Harrogate Roadstone.

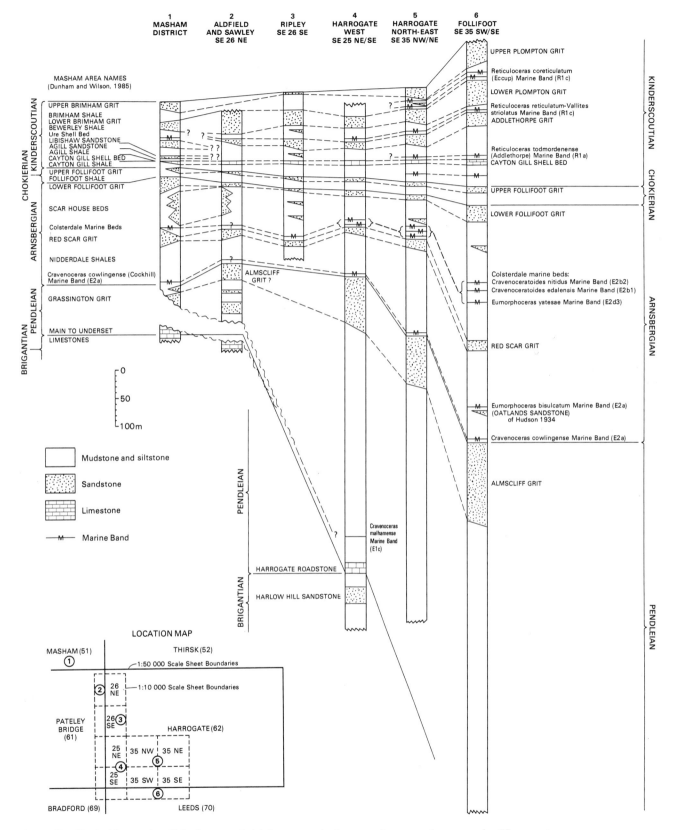

**Figure 7** Early Namurian lithology and thickness variation north to south across the Harrogate district; column 1 refers to the south-east of the Masham district and columns 2 to 6 to 1:10 000 scale maps for the Harrogate district (see inset diagram).

Coalfield, but most details of their sequence and distribution are at present confidential.

## CARBONIFEROUS LIMESTONE

### Pre-Brigantian rocks

Evidence from seismic reflection data suggests that up to 4.5 km of Dinantian rocks occur within the district to the south of the block–basin boundary. However, strata of pre-Brigantian age have been proved only in the Ellenthorpe No 1 Borehole (Figure 6), which was drilled on a seismically defined structural high near Boroughbridge (Falcon and Kent, 1960; Allsop, 1985). Commencing in superficial deposits, it proved a sequence of Triassic and Permian strata down to c.317 m, at which depth it passed into pale grey, dolomitised limestone of Asbian age (Appendix 1). The lowest limestone intersected in the borehole is of Arundian age; it correlates with the Embsay Limestone (Figure 5) of the Skipton Anticline (Hudson and Mitchell, 1937; Stephens et al., 1953). The thickness of the strata between the Arundian limestones and the Asbian limestones (c.390 m) is comparable with that in the Craven Basin although, according to Falcon and Kent (1960, p.33), the sediments in this interval at Ellenthorpe show no evidence of deep-water deposition. Reexamination of the cuttings and core samples shows that many of the limestones include ooliths and other material generated in shallow water, but that turbidity current-generated structures are also present, suggesting reworking of this material from a shelf environment into deeper water.

### Exposed sequence to base of Harrogate Roadstone (Brigantian)

Strata of Brigantian age crop out only in the axial area of the Harrogate Anticline, where they consist of mudstones and siltstones, and the Harlow Hill Sandstone, a correlative of the Pendleside Sandstones. The fine mudstones are basin-floor deposits, usually with a goniatite fauna and laid down in situ; other lithologies were derived from exogenic material, much of it of shallow-water origin, transported into the area by gravity flows and deposited as submarine fans. The Dinantian–Namurian (Lower to Upper Carboniferous) boundary is not marked by a break in sedimentation, nor by any major faunal change other than the incoming of the diagnostic goniatite *Cravenoceras leion*.

### Details

#### BEDS BELOW HARLOW HILL SANDSTONE

This sequence is poorly exposed, the only section being that proved in the Crimple Beck Borehole (Appendix 1 and Institute of Geological Sciences, 1983, p.4). The lowest beds penetrated comprise about 49 m (corrected for dip) of fine-grained sandstone overlain by 35 m of dark grey mudstone containing sandstone laminae and slumped mudstone debris beds. Their fauna includes abundant *Goniatites granosus* of $P_{2a}$ age and they

equate with the middle part of the Lower Bowland Shales of the Craven Basin.

#### HARLOW HILL SANDSTONE

In the Crimple Beck Borehole the Harlow Hill Sandstone (42 m corrected for dip) consists of pale grey, medium- to coarse-grained sandstone, in beds 0.05 to 5.3 m thick, interbedded with micaceous siltstone and fine-grained sandstone. Individual sandstones are commonly massive or slumped at the base, with mudstone clasts, but some are laminated or show cross-stratification; basal loadcasts and flame structures are also present. Many of the massive beds grade upwards into evenly bedded sandstone, commonly topped by 2 to 3 cm of ripple-cross-bedded, fine-grained sandstone. Such characteristics suggest that these strata were deposited from turbidity currents (Leeder, 1982a, p.81).

At the former Harlow Hill Quarry [2870 5411], now filled in, Hudson (1930) recorded the following sequence of sandstone about 15 m thick. 'It is false-bedded, often very thin-bedded and sometimes nodular. Where unweathered it is a greenish grey colour, but it is usually seen weathered to a light brown. Coal fragments occur in the sandstone. Beds of sandy and micaceous mudstone are common and occasionally the sandstone beds are separated by very carbonaceous mudstones. The sandstones are fine-grained, and consist of small rather angular quartz grains with some siliceous cement. Muscovite is common, felspar rare. Thin sandstone bands continue into the shales above. These shales are mainly sulphurous mudstones'. The Harlow Hill Sandstone probably correlates with the last pulse of the Pendleside Sandstones of the Settle district (Figure 5).

#### BEDS BETWEEN THE HARLOW HILL SANDSTONE AND HARROGATE ROADSTONE

In the Crimple Beck Borehole (Appendix 1), the Harlow Hill Sandstone is succeeded by about 25 m (corrected for dip) of mainly dark grey mudstone. This contains sandstone laminae and thin sandstone beds, plus mudstone debris beds. These are overlain by dark grey fossiliferous mudstones with thin crinoidal turbidites. The fauna includes *Sudeticeras* sp. (*adeps* or *newtonense*) and *Sudeticeras* sp. B (Moore, 1950), indicating a probable $P_{2b-c}$ age.

Other details of the sequence are scant. The basal part, formerly exposed near Harlow Hill Quarry [2870 5411], was recorded as sulphurous mudstone; the remainder of the sequence as mudstone, very sulphurous and ferruginous, poorly bedded with thin inconsistent beds of sandstone (Hudson, 1930). At Beckwith Quarry [2808 5236], now largely filled in and formerly called Howe or How Quarry, Edwards (BGS manuscript map) recorded 10.67 m of dark shales below the Harrogate Roadstone. Hudson and Edwards (unpublished BGS manuscript) described the shales as dark and sulphurous, with beds of calcareous sandstone; and they noted that A H Green (unpublished BGS manuscript) recorded '*Spirifer*, *Productus* and a thing like *Saccammina carteri*' in them. These are the beds through which rise the Valley Gardens mineral springs in Harrogate [297 552], their sulphurous nature contributing to the mineral content of the water. They correlate with the topmost part of the Lower Bowland Shales of the Craven Basin.

# MILLSTONE GRIT

## Base of Harrogate Roadstone to top of Almscliff Grit (Pendleian, $E_1$)

Rocks of Pendleian age crop out only around the Harrogate Anticline. The sequence, some 500 m thick, represents the final filling of a basin that had probably existed since Arundian times. Limestone turbidites of deepwater origin (the Harrogate Roadstone), at the base, are overlain by a coarsening-up sequence of delta fan and delta slope deposits with sandstone–siltstone turbidites. The Harrogate Roadstone is the correlative of the Berwick Limestone of the Skipton area. A break in sedimentation, accompanied by folding, has been recognised at the base of the overlying fluviatile Almscliff (Grassington) Grit along the southern edge of the Askrigg Block (Chubb and Hudson, 1925; Dunham and Wilson, 1985; Rowell and Scanlon, 1957; Wilson, 1960). A similar situation occurs along the north-west margin of the Harrogate district (Figure 7, column 2), as proved in the Aldfield and Sawley boreholes. This unconformity appears to die out towards the basin although, it has been postulated that it also occurs within the Craven Basin (Arthurton, 1983). Around the Harrogate Anticline exposures are too restricted to determine whether the break is present there or not.

## Details

### HARROGATE ROADSTONE

The outcrop of the Harrogate Roadstone extends as a narrow strip along the south-eastern side of the Harrogate Anticline from near Shaw Green [264 521] to the Valley Gardens [298 554], where it swings westwards across the north-eastern plunge of the fold. On the north-west flank, it is cut out by the Harrogate Fault for some distance, but reappears [278 539] south-west of Harlow Car Gardens, from where it swings westwards between Beckwithshaw and Shaw Green. In the Crimple Beck Borehole (Appendix 1), the Roadstone is 19 m thick (corrected for dip), consisting throughout of crinoidal limestone turbidites, with individual beds ranging in thickness from less than 0.01 m to over 1.6 m. The lowest 4 m are dark grey and sandy, and the beds average about 0.1 m. The middle 10.5 m consist of pale grey crinoidal beds (average 0.47 m). These show sedimentary structures similar to the thicker beds of the Harlow Hill Sandstone, each of which comprises a basal unbedded part with sporadic slump structures and mudstone clasts, a central bedded part, and an upper ripple-cross-bedded part, capped by a mudstone drape. The top 4.5 m are more thinly bedded (average 0.13 m), darker grey and muddier. The matrix of all the limestones is more or less silicified. The detrital material consists largely of well-graded crinoid columnals, with subordinate crinoidal and bryozoan debris, and some sand-grade quartz.

About 900 m north-north-west of the Crimple Beck Borehole, the Harrogate Roadstone was formerly exposed at Shaw Green Quarries [2690 5267]. Few details of these quarries exist, but Hudson (1930) recorded a thin cherty sandstone beneath the limestone; this accords with the Crimple Beck Borehole section. Hudson and Edwards (unpublished BGS manuscript) give the following description: 'At Shaw Green the Roadstone has been quarried in two openings, separated by a gap which is occupied partly, perhaps wholly, by dark mud-

stone estimated to be 20 to 30 ft (6.10 to 9.15 m) thick. Here crinoidal limestone is interbedded with beds of calcareous siltstone and sandstone'.

The only good exposure of the Roadstone is at Beckwith Quarry [2808 5236], where there is a 6 m section in an unstable cliff below Beckwith House. The sequence here consists of silicified, crinoidal sandy limestone in beds 0.20 to 0.60 m thick, with thin siltstone partings and sporadic mudstone clasts (Plate 1). The rock is cream coloured when fresh, but it weathers to an ochreous rottenstone; characteristic fragments of this occur in the surrounding soil. Hudson (1930) gave a full description of the Harrogate Roadstone at this locality then called How (or Howe) Quarry, Beckwith. He noted that the Roadstone was 45 ft (13.72 m) thick with a sharp basal contact, although Hudson and Edwards (unpublished BGS manuscript) recorded the thickness here as between 35 and 40ft (10.67 to 12.20 m). The following description is largely taken from Hudson (1930), but includes a few additional details from the Hudson and Edwards manuscript: 'The junction with the underlying sulphurous shales and sandstones is quite sharp, but it passes by alternation into the shales above. The rock is mainly a fragmental limestone composed of crinoid ossicles with subordinate fragments of bryozoa and calcareous sponge spicules cemented by chert. When unweathered it is cream coloured with the crinoid ossicles slightly lighter in colour than the rest of the rock. It weathers by solution of the limestone fragments, leaving a vesicular mass of rotten stone. The limestone is very well bedded with the beds usually 6 to 9 ins (0.15 to 0.23 m) thick, though beds up to between 2 ft and 3 ft 6 ins (0.60 and 1.00 m) occur. It is slightly to very false bedded. Thin sandy mudstone lenticles and layers, or shales with thin calcareous sandstones often divide the limestone beds, and often cut sharply downwards into them. Carbonaceous fragments are common in the mudstones. The crinoid ossicles are very small and evenly graded. Large included pebbles of black shale and chert and more rarely black limestone are common in the limestones. The limestones are rarely recognisably cherty though occasionally thin lenticles of grey mottled chert do occur'.

The other notable exposure of the Roadstone was the former Low Harrogate Quarry [2969 5508] on Cold Bath Road. The quarry is mentioned by Fox-Strangways (1874, 1908), but only partially described by Hudleston (1883) who noted (p.427) that 'the crinoidal character is very prevalent and where the soluble matter had been removed it becomes a spongy siliceous encrinite grit'. He also recorded (p.430) that 'In Cold Bath Road Quarry these rocks are seen dipping at a considerable angle in a south-easterly direction'. A newspaper photograph (Acrill Ltd, 1934–35) shows this quarry, which formerly stood on the site of Valley Mount (behind Cold Bath Road). The illustration suggests a thickness of about 14 m of strata in fairly continuous, mainly medium and thick beds with a few thin beds. The apparent dip is about 20°.

The Harrogate Roadstone has not yielded any diagnostic fossils, but its age is fixed as $P_{2c}$–$E_{1a}$ by the mudstone faunas above and below. The equivalent Berwick Limestone (Figure 5) of the Skipton Anticline has yielded *Cravenoceras leion* (Hudson and Mitchell, 1937) suggesting that the Harrogate Roadstone represents the basal beds of the Namurian or straddles the Dinantian–Namurian boundary.

The source of the crinoidal debris probably lay in shallow, high-energy waters near the southern margin of the Askrigg Block. The thickness, composition and purity of individual beds within the Roadstone suggest that it was derived from a single source. The most likely correlative on the Askrigg Block is the Main Limestone, which, on Fountains Fell, is underlain by beds of early $E_1$ age (Arthurton et al., 1988; Hudson, 1941;

**Plate 1**  The Harrogate Roadstone at Howe Quarry [2808 5236], below Beckwith House near Harrogate.

The rock consists of well-bedded, partly silicified, crinoidal, bioclastic limestone containing thin bands and lenses of chert with shaly intercalations (L 1388).

O'Connor, 1964; Rayner, 1953, p.286) and which, farther north (Johnson et al., 1962), contains *Cravenoceras leion*. This was the correlation originally proposed by John Phillips as early as 1865.

BEDS BETWEEN HARROGATE ROADSTONE AND ALMSCLIFF GRIT

Around Harrogate this sequence is about 450 m thick. At its base, about 50 m of mainly dark grey, partly calcareous mudstones with goniatites overlie the Harrogate Roadstone. The basal beds are not exposed, but the lowest 32 m, proved in the Crimple Beck Borehole, contain sporadic thin limestone turbidites and a fauna including *Tumulites pseudobilinguis* of $E_{1b}$ age. Above this sequence the Cravenoceras malhamense Marine Band (basal $E_{1c}$) is exposed at two localities [2546 5178; 2565 5173] in Crimple Beck, where *Posidonia corrugata* and *Cravenoceras malhamense* have been recorded. This highest $E_1$ marine band is here overlain by more than 50 m of dark grey to black, jarositic mudstones with ironstone nodules. They are succeeded by about 350 m of dark grey siltstones with thin, hard, turbiditic sandstones. The highest 250 m of the sequence, up to the Almscliff Grit, are exposed in an almost continuous section (described here from notes supplied by Mr J I Chisholm) in Harlow Car Beck, north of Harlow Car Gardens [2753 5446 to 2780 5421]. Here, the lower 170 m consist of interbedded sandstones and siltstones, including turbidites, interpreted to represent sedimentation in a delta-slope environment. The turbidites comprise graded sandstone beds from 0.1 to 6.0 m thick, with sharp bases; some are coarse grained, with mudstone pebbles. Above, there are 80 m of fine- to medium-grained, mainly parallel laminated sandstones with interbedded siltstones. These are taken to be representative of an upper delta-slope and delta mouth-bar environment and are separated by a 20 m gap from the base of the Almscliff Grit.

## ALMSCLIFF GRIT

The Almscliff Grit ranges in thickness across the area from a few metres in the Masham district to 30 m at Sawley, 100 m around Harrogate and 140 m at Follifoot (Figure 7). It is a coarse- to granule-grained, quartzofeldspathic sandstone containing quartz pebbles throughout; it is thickly bedded and strongly cross-bedded. Around Harrogate, siltstone partings near the top produce separate topographical features above the main outcrop of the grit. On the south-western flank of the Harrogate Anticline, the best section is in Clark Beck, below Yew Tree Lane [297 522]; here, approximately 40 m of coarse-grained, cross-bedded sandstone are present, overlain by about 20 m of flaggy, thin-bedded sandstone and siltstone. On the northern flank of the anticline, the Grit is well exposed in Harlow Car Beck [275 545], at Birk Crag [279 548] and in old quarries to the north-east. One such quarry at Knapping Mount [301 562] exposes 20 m of medium- to coarse-grained, cross-bedded sandstone; this is overlain by 4.4 m of fine-grained, silty sandstone in thick laminated and very thin beds with siltstone interbeds; it is capped by 3 m of medium- to coarse-grained sandstone in medium beds. At Birk Crag [279 548] the steeply dipping, red-stained sandstones form spectacular crags; cross-bedding dip directions here (Chisholm, 1981) indicate palaeoflow from the north-east. The Almscliff Grit equates with the Grassington Grit and marks the first incursion into the district of the coarse, fluviatile, arenaceous facies characteristic of the 'Millstone Grits'.

## Top of Almscliff Grit to top of Upper Follifoot Grit (Arnsbergian, $E_2$, and Chokierian, $H_1$)

These strata range in thickness from a minimum of 165 m on the Askrigg Block to a maximum of 450 m in the Craven Basin (Figure 7). They crop out over a much wider area than the older rocks, being present on both flanks of the Harrogate Anticline and along the axes of the several small anticlines in the region north of the River Nidd. Three persistent sandstones, the Red Scar Grit, the Lower Follifoot Grit and the Upper Follifoot Grit have been mapped (Figure 7); two impersistent sandstones are also present. One, named the Oatlands Sandstone by Hudson (1934), is present between the Almscliff Grit and the Red Scar Grit around Harrogate. The other, between the Red Scar Grit and the Lower Follifoot Grit, is the probable lateral equivalent of the Nesfield Sandstone of the Bradford area (Stephens et al., 1953). Each sandstone marks the top of a complex cyclothem.

## Details

### BEDS BETWEEN ALMSCLIFF GRIT AND RED SCAR GRIT

Unnamed in the Harrogate district, these beds equate with the Cockhill Marine Band and the Nidderdale Shales of Dunham and Wilson (1985, fig. 14). They consist dominantly of siltstone and mudstone, and range in thickness across the district from about 50 m in the north to about 160 m in the south. At the bottom of the sequence, the basal Arnsbergian Cravenoceras cowlingense Marine Band ($E_{2a}$) overlies the Almscliff Grit; it

was proved in a temporary exposure [2745 5464] which yielded *Chaenocardiola footii* and *Cravenoceras cowlingense* (Institute of Geological Sciences, 1971, p.26). It is succeeded by grey silty mudstones, formerly worked for brickmaking in a large quarry [309 539], now filled and landscaped, near Oatlands. The silty mudstone coarsens upwards and is capped by about 6 m of flaggy sandstone, the Oatlands Sandstone; it has been mapped only locally, but may be present throughout the district. The sandstone is seen in a small quarry [3126 5423] north-east of Hookstone Beck and again [2713 5456] 120 m east of Rough Bridge.

Dark grey mudstones above the Oatlands Sandstone in Hookstone Beck [3130 5419] yield a marine fauna including *Posidonia corrugata* and *Eumorphoceras* cf. *ferrimontanum*. Strata on the same horizon in Stone Rings Beck [3051 5275], where the sandstone is not seen, yielded *Dunbarella* sp., *Posidonia?*, *Cravenoceras* sp., *Eumorphoceras erinense* and crinoid columnals. These faunas are diagnostic of the main Eumorphoceras bisulcatum Marine Band ($E_{2a}$). The succeeding grey siltstones and mudstones with thin sandstone bands are capped by the Red Scar Grit; this upper part of the sequence in Coppice Beck [2991 5629] was described by Hudson (1944). South of the district, at Pannal, the Red Scar Grit is faulted against the Almscliff Grit, a coincidence which in the past (Hudson, 1934; Stephens et al., 1953) led to a miscorrelation of strata.

### RED SCAR GRIT

The Red Scar Grit has also been called the Scar Gill or the Hookstone Grit. The unit ranges from between 6 and 10 m thick in the north to around 15 m thick in the south of the district. In the folded and faulted ground north of the River Nidd, the grit occurs in the core of a syncline near Dole Bank [2736 6464]. Although no longer exposed there, previous records of about 4 m of sandstone (Fox-Strangways, unpublished notebook, 1862) suggest that it is thin. On the north flank of the Harrogate Anticline, the grit was formerly exposed in Coppice Beck [2995 5630], where it comprised about 6 m of 'hard buff siliceous grit'(Hudson, 1944, p.235). The rock is still exposed along Oak Beck [2904 5584] and at Oakdale Farm [2810 5526]. Around the eastern closure of the Harrogate Anticline, the Grit is well exposed in a small quarry [3381 5762] on the south bank of the River Nidd north-east of Bilton Hall; here, it is 15 m thick and dips eastwards. The same bed is again seen in the river 900 m upstream [3295 5775], where it dips to the north-west. On the south flank of the Harrogate Anticline, the Red Scar Grit is well exposed in Longlands Quarry [3145 5335]. The sequence here comprises over 14 m of medium- to coarse-grained, cross-bedded sandstone. Other exposures in the vicinity include Hookstone Quarries [3181 5393], which are largely obscured by tipped material, but where 6 m of medium-grained sandstone are still exposed.

### BEDS BETWEEN RED SCAR GRIT AND LOWER FOLLIFOOT GRIT

These beds, equivalent to the Colsterdale Marine Beds and Scar House Beds of Dunham and Wilson (1985, fig. 14), range in thickness from 60 to 205 m. From the west of Harrogate the thickness increases dramatically southwards and slightly northwards (Figure 7). The interval is composed dominantly of siltstone and mudstone with a minor sandstone unit about 100 m above its base near Harrogate. The basal part of the sequence comprises fossiliferous mudstones, the Colsterdale Marine Beds of $E_{2b}$ age, which are at least 30 m thick. Over 52 m were formerly exposed in Coppice Beck; they were described by Hudson (1944). The basal beds comprised the Eumorphoceras yatesae Marine Band ($E_{2a3}$) with a basal leaf rich in the bivalve *Leiopteria* sp.. The marine band is about 3 m thick and contained *Eumorphoceras yatesae*. The base of the succeeding Cravenoceratoides edalensis Marine Band ($E_{2b1}$) occurred 13.4 m above that of *E. yatesae*; the marine band was about 3.6 m thick and the upper leaf contained bullions yielding *Cravenoceratoides edalensis*. Details of the faunas above $E_{2b1}$ are poorly known; however, the Nitidus Limestone (0.3 m thick), which lies within the $E_{2b2}$ Cravenoceratoides nitidus Marine Band, was 17.68 m above the $E_{2b1}$ Marine Band and separated from it by dark marine mudstones. The Nitidus Limestone was overlain by 14.5 m of unfossiliferous dark mudstone containing small nodules.

Equivalent beds to these were previously worked in the Brick Pit at Stonefall [331 548], which was reported to be 40 m deep when it closed. Hudson et al. (1938, p.370) noted that the shales here contained nodular limestone masses with *Cravenoceras* and *Anthracoceras*; they also recorded purple-red staining (Permian ?) in the upper 25 ft (7.6 m) of the section. Unpublished BGS palaeontological records contain the following section for the pit (recorded in about 1930): 'About 70 ft (21.34 m) of dark grey shaly mudstone with beds of siltstone and silty sandstone with septarian nodules; upper half stained as at Killinghall'. Sir James Stubblefield determined the following species from the locality: orthocone nautiloid, *Anthracoceras paucilobum*, *Cravenoceratoides nitidus*, *Ct. nititoides*, *Dimorphoceras* sp. and *Eumorphoceras* sp. The presence of *Ct. nititoides* indicates an $E_{2b3}$ level, but the material was found loose and the band was not located on the quarry face.

The Colsterdale Marine Beds are now seen in Crimple Beck between Pannal Bridge [3074 5166] and Almsford Bridge [311 523]; here, in the river banks, there are intermittent sections of dark grey mudstone, calcareous siltstone and shaly limestone with a fauna dominated by bivalves and goniatites including: *Ianthinopsis* sp., *Posidonia corrugata*, *P. corrugata gigantea*, *Selenimyalina variabilis* (common), orthocone nautiloids, *Anthracoceras* sp. (common), dimorphoceratids, *Cravenoceras* sp., *Cravenoceratoides* sp. and *Eumorphoceras* sp.; these faunas probably lie within $E_{2b2}$. In the north, on the Hardgate Anticline [2662 6288], similar fossiliferous mudstones are also exposed. Above Coppice Beck [300 562] the minor sandstone which caps the cyclothem is partly visible. This sandstone, about 100 m above the Red Scar Grit, is the probable lateral equivalent of the Nesfield Sandstone of the Bradford area (Stephens et al., 1953) and part of the Scar House Beds of Dunham and Wilson (1985, fig. 14). Strata of the succeeding cycle, up to the Lower Follifoot Grit, are not exposed.

Dunham and Wilson (1985, fig. 14) suggest that, southwards across the Askrigg Block, the sandstones of the Scar House Beds become finer grained and include more siltstone in the sequence. This accords with available information in the Harrogate district, and with evidence from farther south where the BGS Harewood Borehole [3220 4410] terminated in the $E_{2a3}$ Eumorphoceras yatesae Marine Band, overlying the Red Scar Grit. This marine band is overlain by a complete $E_{2b1-2}$ and $E_{2c1}$ sequence of marine bands, without any intervening sandstones.

## LOWER FOLLIFOOT GRIT

The Lower Follifoot Grit ranges in thickness from as little as 3 m in the north to 15 m in the south. In the north it crops out in the cores of small anticlines. It is poorly exposed in the bed of the beck [2791 6473] near Markington; here, the rock is a very coarse-grained, quartzose, silica-cemented sandstone. Exposures just west of the district show the rock to be locally ferruginous and ganisteroid. Around Harrogate, the Lower Follifoot Grit is at least 12 m thick, evenly bedded and fine to medi-

um grained; it includes a thin, dirty coal and seatearth 6m below the top. On the northern flank of the Harrogate Anticline, the grit forms a feature south of Saltergate Beck [277 566]. The grit is seen in the Nidd gorge west of Conyngham Hall [3421 5741], where it dips eastwards at 25 to 30° and forms a small escarpment projecting above the general level of the sub-Permian unconformity; it forms a low ridge against which the basal Permian sediments are banked. In the disused railway cutting near Home Farm, Rudding Park [3235 5265], the grit is well exposed, the section here being:

|  | *Thickness* m |
| --- | --- |
| Sandstone, fine-grained, flaggy, with sandy shale partings | 6.2 |
| Coal, dirty | 0.05 |
| Seatclay, sandy | 0.6 |
| Sandstone, fine- to medium-grained, evenly bedded | 4.5 |

From here along the south flank of the Harrogate Anticline the grit is not exposed, but forms a distinct feature parallel to that of the Upper Follifoot Grit.

## BEDS BETWEEN LOWER AND UPPER FOLLIFOOT GRITS

The sequence between the Lower and Upper Follifoot grits (Follifoot Shale of Dunham and Wilson, 1985, fig. 14) is poorly exposed. It comprises mudstones with subordinate sandstones and ranges in thickness from 12 to 22 m. In Cayton Gill, a stream section [2860 6258] showed the mudstones to contain sideritic ironstone nodules and the sandstones to be fine to medium grained in thin beds. In the pipeline trench near Bishop Thornton [264 633], similar lithologies were recorded from this interval by C G Godwin (unpublished notebook, 1969–71). Throughout the remainder of the district, these mudstones are known only from boreholes. Outside the district, the interval has yielded goniatites of Chokierian and latest Arnsbergian age (Hudson, 1939; Wilson, 1960, 1977).

## UPPER FOLLIFOOT GRIT

This sandstone sequence is poorly exposed in the Harrogate district, where it ranges in thickness from 2 to 9 m in the north to around 12 m in the south. Exposures in the north are insignificant and the grit is known only as a ridge-like feature north of Harrogate, to the south of Saltergate Beck [277 566]. The grit is seen in the Nidd gorge near Conyngham Hall [3422 5746], where it dips eastwards at 25 to 30°. Like the Lower Follifoot Grit it forms a small escarpment projecting above the general level of the sub-Permian unconformity, and the basal Permian sediments are banked against it. This locality exposes 7 m of fine- and medium-grained sandstone with subordinate siltstone partings and a thick bed of siliceous ganister with rootlets near the middle of the unit. East of the anticline, the grit is proved in the Farnham Borehole (Burgess and Cooper, 1980b) where the top 3 m were penetrated; these beds comprise fine- and medium-grained sandstone with ripple cross-lamination and zones of intense bioturbation. South of Harrogate the Upper Follifoot Grit is not exposed, but forms a ridge through Rudding Park and Follifoot Ridge Farm.

A non-sequence, representing the whole of Alportian ($H_2$) time, is believed to be present in this sequence, between the top of the Upper Follifoot Grit and the overlying marine bands (Ramsbottom, 1977, p.273). There is no angular discordance, the gap in sequence being recognised only by the absence of the appropriate goniatite faunas (Figure 5). The evidence is derived largely from sequences in adjacent districts (e.g. Wilson, 1977, fig. 2).

## Top of Upper Follifoot Grit to top of Upper Plompton Grit (Kinderscoutian, $R_1$)

The pattern of cyclic sedimentation established in the Arnsbergian continued during the Kinderscoutian, this part of the sequence being equivalent to the Kinderscout Grit Group of the Bradford area. The lowest beds, of $R_{1a}$ age, are mainly marine mudstones, but include the Cayton Gill Shell Bed recognised by Fox-Strangways (1874) and Tute (1886), which is a shelly sandy siltstone. This bed forms a marker horizon, which is invaluable in elucidating the stratigraphy and structure of the area, as it is easily recognised from its characteristic mode of weathering into platy slabs containing abundant decalcified shells.

The succeeding strata include several coarsening-upwards cycles, each capped by sandstone. The three most persistent are the Addlethorpe, Lower Plompton and Upper Plompton grits. A cored borehole drilled at Stockeld (Appendix 1), near Wetherby, in the adjoining Leeds district(Sheet 70), proved this sequence. A reappraisal by Dr W H C Ramsbottom of the faunas from the borehole (Burgess and Cooper, 1980b) showed that the Addlethorpe Grit is of $R_{1b}$ age, being overlain by the Reticuloceras reticulatum–Vallites striolatus Marine Band ($R_{1c}{}^1$), and that the Reticuloceras reticulatum–Reticuloceras coreticulatum Marine Band ($R_{1c}{}^4$ and known locally as the Eccup Marine Band) lies between the Plompton Grits.

In the Leeds district, the three sandstones are named the Addlethorpe, Caley Crags and Bramhope grits respectively (Edwards et al., 1950). However, the last two names are derived from the Bradford (Sheet 69) district, where the presence of the basal $R_{1c}$ Reticuloceras reticulatum–Vallites striolatus Marine Band above the Caley Crags Grit, and the Eccup Marine Band between the Addingham Edge and Bramhope grits (Stephens et al., 1953, pp.33–46, fig. 8) shows that the correct correlations are: Caley Crags Grit (Bradford) equals Addlethorpe Grit (Leeds and Harrogate); Addingham Edge and Bramhope grits (Bradford) equal Lower and Upper Plompton grits (Harrogate). The 'Addlethorpe Grit' of Bradford and the 'Caley Crags Grit' of Leeds are incorrectly named (Figure 5).

The correlation of the Otley Shell Bed, underlying the Caley Crags Grit (Stephens et al., 1953, pp.39–41), with the Ure Shell Bed ($R_{1b}$), underlying the Lower Brimham Grit near Pateley Bridge and Masham, by Dr A A Wilson (Rayner and Hemingway, 1974, fig. 30) could indicate a likely correlation of the Lower Brimham Grit with the Caley Crags Grit, and hence with the Addlethorpe Grit as suggested by Burgess and Cooper (1980b), although faunas of $R_{1b}$ age are locally absent in the Harrogate and Wetherby areas, probably as a result of erosion at the base of the Addlethorpe Grit. However, the correlation of the two shell beds is tenuous (Wilson and Thomson, 1965; Wilson personal communcation, 1990) and alternative correlations between the Masham and Harrogate districts are possible. In view of the known tectonic instability of the southern margin of the Askrigg Block, resolution of these correlation problems must await the mapping of the intervening ground around Pateley Bridge.

### Details

BEDS BETWEEN UPPER FOLLIFOOT GRIT AND CAYTON GILL SHELL BED

This sequence, underlying the Cayton Gill Shell Bed and equivalent to the Cayton Gill Shale of Dunham and Wilson (1985, fig. 14), is composed of micaceous siltstones and fossiliferous mudstones, with thin, bioturbated shelly sandstones. Borehole evidence indicates a thickness of 30–45 m, but good surface sections are generally lacking. In a borehole [2639 5719] at Saltergate Hill, 42 m of shale were recorded. Nearby, a section [2782 5665] on the north side of Saltergate Beck, 150 m south of Killinghall Moor Farm, exposes 1.5 m of hard, flaggy, dark grey siltstone with ribs of pale grey siliceous sandstone; the sandstone is bioturbated and contains calcareous shelly debris. Downstream from this exposure for 130 m there are small exposures of grey siltstone. Eastwards along the stream, beyond a minor fault, there are several small outcrops of contorted grey siltstone with thin sandstone beds. Mudstones and siltstones in this interval are poorly exposed on the south bank of the River Nidd [3424 5745] north of Conyngham Hall, but elsewhere they are not exposed. In the Farnham Borehole these beds are 32.5 m thick; they are mainly micaceous siltstones and fossiliferous mudstones, with several thin beds of bioturbated, shelly, calcareous sandstone. The marine beds have yielded the following fauna: *Lingula mytilloides* (common), *Orbiculoidea cincta*, *O. nitida*, *Productus carbonarius* (common), *Rugosochonetes laguessianus*, *Coleolus* sp., *Euphemites* sp., *Retispira* sp., *Caneyella squamula*, *Parallelodon* cf. *regularis*, *Streblochondria* sp., *Catastroboceras kilbridense*, ostracods, crinoid columnals, fish scales and debris.

CAYTON GILL SHELL BED

The Cayton Gill Shell Bed is a distinctive marker horizon throughout the district. It is present in the cores of small anticlines north of the River Nidd and has a narrow, mainly drift-covered outcrop on both flanks of the Harrogate Anticline. It is harder than the surrounding mudstones and generally forms a marked feature, covered by the characteristic weathered slabs of shelly, siliceous mudstone, siltstone and sandstone! The type section is on the west side of Cayton Gill [2774 6324] (see below), where it has a thickness of 3.54 m; nearby sections are similar, with 3.25 m of shell bed exposed [2771 6326] adjacent to the type section.

|  | *Thickness* m |
| --- | --- |
| Top of section, no exposure | |
| Mudstone, very weathered, orange-brown, soft and clayey | 0.70 |
| *Top of Shell Bed* | |
| Sandstone, dark beige, medium-grained, fine-grained at base; thinly bedded, calcareous in parts, with fossil fragments. The upper part of the unit is very thinly bedded, very fine-grained sandstone, with siltstone and mudstone partings | 0.90 |
| Shell bed of very calcareous mudstone, light grey, hard and roughly banded in thin to medium beds. The beds are packed with complete and fragmental fossils including abundant crinoid | |

| | |
|---|---|
| columnals and brachiopods | 1.25 |
| Sandstone, beige, medium-grained, silty and calcareous, medium and thickly bedded; splitting into medium beds in places | 1.18 |
| Sandstone, medium beige-grey, very fine- to fine-grained, very thin to thickly laminated bedding, with partings of dark grey mudstone | 0.21 |

*Bottom of Shell Bed*

| | |
|---|---|
| Siltstone, dark grey-beige, slightly sandy towards top, very thin to thick laminated bedding. The rock passes into soft mudstone towards its base | 0.34 |
| No exposure | 1.35 |
| Siltstone, dark grey and dark beige, hard, slightly banded and slightly bioturbated in places; three small exposures in brook | about 2.0 |

Another exposure, on the east side of Cayton Gill [2867 6278], 220 m west of Cayton Gill Farm, comprises 2.6 m of calcareous sandstone and sandy limestone; the beds are grey, weathering brown and orange, thin- and medium-bedded, with abundant crinoid and brachiopod fragments. On the western edge of the district the shell bed was recorded in several places along the Ure Valley water main trench to the west of Cayton Gill (Godwin, 1973). In the Farnham Borehole [3469 5996] the Cayton Gill Shell Bed is 7.64 m thick. It is represented by 5.96 m of fine-grained sandstone with carbonaceous debris and a siliceous cement towards the top, overlain by 0.15 m of shelly, recrystallised limestone with cone-in-cone stuctures, followed by 0.46 m of dark grey silty mudstone with shelly debris; the sequence is capped by 0.27 m of muddy, crinoidal and shelly limestone. Just west of the district, at Bungalow Farm [2624 5968], the shell bed is well exposed. Here it comprises 2.2 m of shelly, calcareous sandstone overlain by 0.05 m of grey siltstone and 1.75 m of flaggy, siliceous limestone with cherty laminae. Farther south, around Harrogate, the shell bed is thicker (up to 12 m), sandier and cherty, but exposures are small and scarce. In Saltergate Beck [2811 5658], 1.0 m of flaggy, shelly sandstone with a siliceous cement is seen resting on grey siltstone. South of Harrogate, the shell bed was formerly exposed in numerous small quarries around Rudding Park and Rudfarlington Farm. Hudson (1934, p.124) noted that 'South of Harrogate, the Cayton Gill Beds are very different from the fossiliferous shell beds north and west of Harrogate. At Rudding Park and Spofforth Haggs where they are well exposed they are represented by 30 or 40 ft (9.15 or 12.20 m) of beds of light coloured chert with occasional brachiopod casts and with bands of marly mudstone. Thin sections show the chert to be chert-cemented siliceous silt.'

The Cayton Gill Shell Bed is the most extensive and species-rich of several benthonic fauna-bearing beds in strata of $R_1$ age in central Yorkshire. Within this district and farther west it has yielded abundant brachiopods and bivalves, with small numbers of bryozoa, gastropods and corals. Some of the brachiopods have been described by George (1932) and Legrand-Blain (1985). These rich $R_1$ benthonic faunas are especially significant because they represent a rare presence in Britain of shelly faunas closely postdating the mid-Carboniferous (approximately Mississippian–Pennsylvanian, or approximately Serpukhovian–Bashkirian) boundary (Harland et al., 1989). Species identified from the district include: *Hyalostelia smithii, Fenestella* cf. *plebeia, Rhombopora lepidodendroides, Lingula mytilloides, Orbiculoidea nitida, Choristites* sp., *Crurithyris* sp., *Derbyia* sp., *Dictyoclostus pinguis, Eomarginifera* sp., *Juresania* sp., *Linoproductus* cf. *postovatus, Marginifera* sp., *Martinia* sp., *Orthotetes gigantea, Productus carbonarius* (abundant), *Pugilis* cf. *senilis, Punctospirifer northi, Rugosochonetes hindi* (abundant), *R. laguessianus,*

*Schizophoria hudsoni* (abundant), *Spiriferellina* cf. *perplicata, S.* aff. *insculpta, Euphemites* sp., *Retispira* sp., *Aviculopecten interstitialis, A. mutica, Caneyella* sp., *Edmondia* sp., *Parallelodon* cf. *regularis, Streblochondria* sp. and *Pseudocatastroboceras rawsoni.* The fauna is comparatively rich for the British Kinderscoutian and in the area to the west the bed is a rare north-west European source of mid-Carboniferous corals (Wilson, 1977).

### BEDS BETWEEN CAYTON GILL SHELL BED AND ADDLETHORPE GRIT

This interval is very poorly exposed. It is dominantly composed of mudstone and siltstone, and increases in thickness across the area from 20 m in the north to 60 m in the south. Locally, in the vicinity of Cayton Gill, an impersistent sandstone unit is also present in the middle of the sequence. The only detailed section through the interval was provided by the Farnham Borehole [3469 5996] (Figure 8) which proved 31 m of beds representing five cyclic units. The basal part of the lowest unit, above the Cayton Gill Shell Bed, comprises 6.0 m of fossiliferous marine mudstones including a 0.27 m-thick bed of muddy, crinoidal limestone. This is succeeded by a seatearth and thin sandstone totalling 0.64 m in thickness. The second cycle is 16 m thick and starts with a medium bed of shelly crinoidal limestone. This is followed by 12 m of dark grey mudstone which has yielded a fauna including *Rotophyllum* sp., *Sphenothallus* sp. *Crurithris* sp., *Lingula mytilloides, Orbiculoidea* sp., *Productus carbonarius, Rugosochonetes* sp., *Schizophoria* sp., *Coleolus* sp. *Bellerophon* sp., *Euphemites* sp., *Anthraconeilo laevirostrum* (common), *Metacoceras* sp., *Dimorphoceras* s.l. sp., *Reticuloceras paucicrenulatum,* ostracods, crinoid columnals, fish scales and debris. The fauna, which was also found west of the Harrogate district in the River Washburn (Wilson, 1977), is indicative of the *Reticuloceras todmordenense* Marine Band, and is equivalent to the Addlethorpe Marine Band of Edwards et al. (1950); the cycle is capped by silty, bioturbated sandstone. The third, fourth and fifth cycles are similar to the second, but thinner (3.42 m, 1.47 m and 3.50 m thick respectively). The top cycle is truncated by erosion at the base of the Addlethorpe Grit. The sandstones in the second and third cycles may be equivalent to the Agill and Libishaw sandstones of the Masham district (Wilson, 1960) and the Libishaw Sandstone of the Fewston area (Wilson, 1977).

The highest beds in this sequence, immediately underlying the Addlethorpe Grit, are exposed in a cliff section [3431 5746] below Conyngham Hall where c.8 m of interlaminated sandstones and siltstones may be seen. The 5 m of fine-grained, flaggy and cross-bedded sandstone under the High Bridge at Knaresborough [3452 5711] may also lie at this level.

### ADDLETHORPE GRIT

The Addlethorpe Grit is generally a coarse-grained, quartzo-feldspathic, thickly bedded sandstone, with cross-bedded units of up to 6 m, although locally, around Follifoot, it becomes thinly bedded, flaggy and micaceous. The unit ranges in thickness from 8 to 22 m and equates with the Caley Crags Grit of Bradford (Figures 5 and 7); a correlation with the Libishaw Sandstone or the Lower Brimham Grit of Masham (Dunham and Wilson, 1985) is also possible.

In the north the Addlethorpe Grit is inferred to cross the Skell valley in the vicinity of Fountains Abbey [2717 6821]; farther south, the grit forms features on either side of Cayton Gill. Nearby it is well exposed in a quarry at Scarah [278 616] where the 15.2 m section at the top of the unit is:

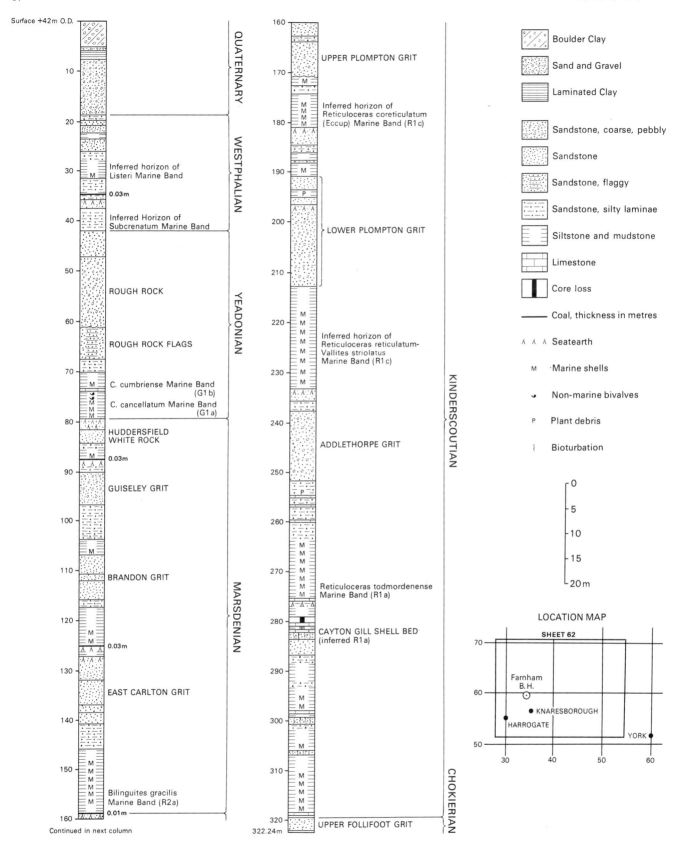

**Figure 8**   Simplified graphic log of the Farnham Borehole [3469 5996], modified from Burgess and Cooper (1980b).

| | Thickness m |
|---|---:|
| Sandstone, silty and siltstone, brown, fine-grained, thinly and very thinly bedded | 1.3 |
| Mudstone and siltstone, dark grey, very thinly bedded, with sandstone partings | 0.95 |
| Siltstone and silty fine-grained sandstone, grey and brown, laminated and very thinly bedded | 2.25 |
| Sandstone, brown, coarse and very coarse-grained, feldspathic, thick and very thick-bedded with cross-bedding | 2.9 |
| Sandstone, brown, coarse- to granule-grained, feldspathic, thick- and very thick-bedded, with cross-bedding | 4.1 |
| Sandstone, brown, very coarse- to granule-grained, feldspathic, with small quartz pebbles | 3.7 |

Another section at Kettle Spring [2692 6241] shows 12.1 m of similar coarse-grained strata, the lowest bed being massive and 4 to 5 m thick. There are a number of small exposures of coarse-grained, quartzofeldspathic sandstone in Hollybank Wood [2702 5986], on the north bank of the River Nidd. Farther south the grit is concealed by drift for 2 km, but is again well exposed in quarries on either side of Oak Beck. Here the thickest section is that seen in the more easterly of the Knox Quarries [2970 5705], where about 18 m of thickly bedded, partly cross-bedded, coarse-grained, quartzofeldspathic sandstone are present; a nearby borehole [2995 5710] proved the sandstone to be 18.9 m thick. Other quarries in the vicinity [2837 5716; 2930 5712; 2955 5705] expose shorter sections of similar lithology. Farther east, the outcrop is mainly concealed by drift and made ground, but small exposures in Roche Avenue [3116 5722] and in the disused railway cutting [3163 5735] are probably also in this sandstone.

In the River Nidd gorge, east of Scotton Banks Hospital [3295 5805], at least 15 m of red-stained, thickly bedded, cross-bedded and coarse-grained, quartzofeldspathic sandstone are exposed on both banks of the river; this is assigned to the Addlethorpe Grit. About 200 m north-east of Conyngham Hall [at 3439 5752], a small thickness of coarse sandstone underlies the Lower Magnesian Limestone; it has a sharp transgressive base and may also belong to this grit. Farther south, in the River Nidd gorge, between High Bridge [3450 5712] and Grimbald Bridge [3618 5618], similar beds are also exposed beneath the Permian unconformity. It is not certain to which grit they belong, but the coarse-grained, quartzofeldspathic, cross-bedded sandstones exposed in the river by the weir [3603 5585] and in the cliff section on the opposite bank [3608 5683] are tentatively assigned to the Addlethorpe Grit, though they may belong to the succeeding Plompton grits.

In the Farnham Borehole (Figure 8) to the east of Knaresborough the Addlethorpe Grit is 14 m thick, with a sharp erosive base. It is dominantly very coarse- to granule-grained, quartzofeldspathic sandstone with abundant quartz pebbles and large pink feldspar grains up to 1 cm across. In places it is micaceous, and large-scale cross-bedding is the dominant sedimentary structure.

South of the River Nidd, the grit forms a feature above Wingate Farm; a trench [3489 5375] south of the farm exposed a section in flaggy sandstone. Beyond Crimple Beck, the grit is only poorly seen in the disused railway cutting of Spofforth Moor [3335 5199; 3365 5185], where there are discontinuous exposures of flaggy, fine- to medium-grained sandstone with plant debris.

Just to the south of the Harrogate district, the Stockeld Borehole [3803 4945] (Appendix 1) proved the full thickness of the Addlethorpe Grit to be only 6.7 m, comprising fine- to coarse-grained flaggy sandstone, with slumped beds. This confirms that the fluviatile grit facies seen in the Harrogate district thins southwards as indicated by Edwards et al. (1950, p.7).

## BEDS BETWEEN ADDLETHORPE GRIT AND LOWER PLOMPTON GRIT

The beds overlying the Addlethorpe Grit, of basal $R_{1c}$ age, are generally very poorly exposed, with scattered small outcrops showing only a metre or so of mudstone or siltstone. In the north of the district, where the sequence is between 22 and 27 m thick, a thin sandstone unit is present in the mainly siltstone sequence. In the Nidd gorge, west of Scotton Banks Hospital [3318 5822], the thickness is about 15 m. The complete sequence was proved in the Farnham Borehole (Figure 8) where the beds were 20 m thick. They comprise mainly mudstone with an abundant brachiopod and gastropod fauna, plus interbedded siltstones and fine-grained sandstones. These marine beds were formerly exposed in Grange Brick Works [2865 5770], near Killinghall. The section is now largely obscured, but a section including data from Hudson et al. (1938, p.354) and unpublished BGS records is:

| | Thickness m |
|---|---:|
| Grey clay with pebbles (till) | 0.5 |
| Seatearth siltstone, pale grey | 1.0 |
| Sandstone, medium- to coarse-grained, cross-bedded, red-stained in top 0.3 m | 1.0–3.2 |
| Siltstone, dark grey, micaceous, with sporadic 0.05 to 0.10 m sandstone beds | c.10.0 |
| Mudstone, black and shaly, with concretions | c.4.0 |
| Mudstone, black and shaly, fossiliferous | 0.6 |
| Concretionary bed, fossiliferous | 0.2 |
| Mudstone, black and shaly, poorly fossiliferous | c.4.3 |
| Ripple-marked surface on top of underlying grit | |

The sandstone at the top of the section closely underlies the Lower Plompton Grit. Fossils in the mudstones, both in situ and from numerous loose concretions, include the following: *Lingula mytilloides* (common), *Orbiculoidea nitida*, *Neochonetes* sp., *Productus* cf. *carbonarius*, *Glabrocingulum armstrongi*, *Hemizyga?* sp., *Ianthiniopsis* sp., *Microptychis* aff. *propensum* (common), *Retispira* sp., *Anthraconeilo laevirostrum* (common), *A. luciniformis*, *Edmondia* cf. *senilis*, *Myalina pernoides* (common), *Nuculopsis gibbosa*, *Phestia attenuata*, *Prothyris?*, *Sanguinolites* sp. nov., *Schizodus portlocki*, *Brachycycloceras* sp., *Parametacoceras pulcher*, *Anthracoceras* sp. s.l., *Reticuloceras reticulatum.*, and *Vallites* sp.; Hudson et al. (1938, p.354) also recorded *Homoceras striolatum* (now *Vallites striolatus*), further confirming the $R_{1c}$ age of these beds.

In the Farnham borehole this interval yielded the following fauna: *Sphenothallus* sp., *Lingula mytilloides* (common), *Orbiculoidea cincta*, *O. nitida*, orthotetoid, *Productus carbonarius* (common), *Rugosochonetes* cf. *hindi*, *Bellerophon* sp., *Euphemites* sp., *Ianthiniopsis* sp., *Retispira* sp., *Anthraconeilo laevirostrum* (common), *Aviculopecten* cf. *dorlodoti*, *Caneyella squamula*, *Edmondia* cf. *senilis*, *Phestia attenuata*, *P. brevirostris*, *Schizodus* sp., orthocone nautiloid, *Dimorphoceras s.l.* sp., crinoid columnals, fish scales and debris.

Just south of the district, the Stockeld Borehole (Appendix 1) proved this interval to be 40.06 m thick, composed dominantly of siltstone and mudstone with subordinate sandstone beds. At the base of the unit, fossiliferous marine mudstones yielded a fauna indicative of one of the Reticuloceras reticulatum marine bands, possibly that of $R_{1c1}$.

## LOWER PLOMPTON GRIT

The Lower Plompton Grit ranges in thickness from about 30 m in the north of the district, to 20 m over the Harrogate Anticline, and 45 m in the south. It is a medium- to very coarse-grained, locally pebbly, cross-bedded, feldspathic sandstone, flaggy and micaceous at the top, with mudstone partings. In the north, the grit forms conspicuous scars on either side of the River Skell around Fountains Abbey [275 683]. The quarried cliff north of the abbey [2755 6837] exposes up to 7.3 m of cross-bedded, fine- to coarse-grained, pebbly, quartzofeldspathic sandstone. Between there and Ripley, it has been mapped from features; exposures are few. Small quarries at Scarah Moor Farm [2800 6231] and near Ripley Castle [2825 6032] expose sandstones resembling those at Fountains Abbey. The upper part of the sequence in the Ripley area is composed of fine- to medium-grained, thin- to medium-bedded, slightly micaceous sandstones with intervening mudstones. A 10 m section is seen in Killinghall Quarry [290 596], where cross-bedded, coarse-grained, quartzofeldspathic sandstone is visible. Two quarries near Knox [290 578; 289 579] each expose a single 6 m-thick cross-bedded unit and, farther east, a 14 m section is seen in the River Nidd [3268 5827]. Farther south, the sandstone is seen again at Brown Hill Quarry [3504 5355]. This quarry, 15 m deep, exposes 12 m of very coarse- to granule-grained, quartzofeldspathic sandstone with quartz pebbles, in thick and very thick beds with large-scale cross-bedding. The most southerly section is in the disused Spofforth railway cutting [348 516], where over 30 m of red-stained, cross-bedded sandstones, similar to those described above, are exposed. South of the district, the Stockeld Borehole [3803 4945] (Appendix 1) proves the grit to be 21.8 m thick.

## BEDS BETWEEN LOWER AND UPPER PLOMPTON GRITS (INCLUDING BILTON COAL)

The sequence between the Lower and Upper Plompton Grits ranges in thickness from 20 to 30 m. It is very poorly exposed and dominantly siltstone and mudstone, but it also includes 5 m of coarse, quartzofeldspathic sandstone (Sandstone R of Burgess and Cooper, 1980b) which forms a strong feature on Bilton Banks [3254 5811 to 3240 5830]. This sandstone underlies marine beds inferred to be the Reticuloceras coreticulatum (Eccup) Marine Band. In the Farnham Borehole, this interval yielded *Sphenothallus* sp., *Lingula mytilloides* (common), *Orbiculoidea cincta*, *Retispira* cf. *undata*, *Anthraconeilo laevirostrum*, fish scales and debris. Beneath these marine beds, the locally developed Bilton coals occur (Fox-Strangways, 1908, p.6). The upper, principal seam, 0.97 m thick, is separated from the lower (0 to 0.84 m) by 2.7 to 3.6 m of 'shale'. The old workings extend 500 m westwards: five abandoned adits are visible on Bilton Banks and at least 18 shafts (the deepest to 55 m) are located north-east of Bilton Village Farm [319 578] (see 1:10 000-scale BGS map SE 35 NW for the locations of the shafts). The coals were worked until the 1920s, being brought by tramway to Bilton Lime Quarry [3215 5770] and used to fire the kilns. The coals appear to be a local development; they thin out eastwards. Throughout the southern part of the district information about this part of the sequence is scant. A little farther south, the Stockeld Borehole (Appendix 1) proved 28.66 m of mainly shaly measures, with the Eccup Marine Band at the base, above the Lower Plompton Grit.

## UPPER PLOMPTON GRIT

The Upper Plompton Grit, a medium- to coarse-grained, quartzofeldspathic sandstone, 11 to 40 m thick, has not been mapped north of the River Nidd. It is seen in old quarries south of Scotton where, in the most westerly of them [3255 5835], a vertical cliff section shows 16.8 m of micaceous, medium- and coarse-grained, cross-bedded sandstone. In the Farnham Borehole (Figure 8) the Upper Plompton Grit is only 11.66 m thick and has a sharp and erosive base. The sandstone is cross-bedded, mainly very coarse- to granule-grained with layers of small pebbles. The top 0.3 m of the unit is a fine-grained, siliceous sandstone seatearth with carbonaceous rootlets.

In the Nidd gorge, east of Low Bridge, red-stained, quartzofeldspathic sandstones are well exposed beneath the Permian unconformity in the old quarries at Abbey Crags [3550 5593 to 3578 5570] (Plate 2) and again at Grimbald Crag [3610 5577], where over 28 m of cross-bedded sandstone may be seen. These sandstones are all very coarse- to granule-grained, with quartz pebbles and abundant feldspar grains; their petrological characteristics and heavy mineral content are described by Gilligan (1920). These outcrops probably belong to the Upper Plompton Grit, but because of the lack of stratigraphical control it is possible that they could belong to the Lower Plompton Grit. At these localities the sandstones are reddened and weathered in a similar manner to those at Plompton Park (see below), but here they are unconformably buried by the Lower Magnesian Limestone, suggesting that the reddening generally may be of Permian origin.

Farther south, crags and stacks of thickly cross-bedded, coarse-grained pebbly sandstone are magnificently exposed (Plate 3) in Plompton Park [356 537], the type locality, and in fields west and south of Bramham Hall [358 527]. Here, the characteristic reddening and tor-weathering are well developed (Palmer and Radley, 1961). At Plompton Park, the lakeside crags expose up to 9 m of beds, with a further 4 m of strata reported to be present below the water and visible on occasions when the lake is drained. Lithologically, the sandstones are identical to those exposed in Knaresborough gorge and described above. The same grit is also seen in the disused railway cutting at Haggs Bridge [3540 5147] and at Newsome Bridge Quarry [3792 5146] (Plate 4), near North Deighton. The Stockeld Borehole [3803 4945] (Appendix 1), just south of the Harrogate district, started in the grit and proved the lower 8.53 m of it.

## Top of Upper Plompton Grit to top of Rough Rock (Marsdenian, $R_2$, and Yeadonian, $G_1$)

Late Namurian strata crop out over an area of only 10 km$^2$, on either side of the River Nidd, between Bilton and Nidd in the west and Scriven and Farnham in the east. This part of the sequence is almost unexposed, but was proved in the Farnham Borehole (Figure 8; Burgess and Cooper, 1980b). The sequence is closely comparable to, though thinner than that in the Leeds–Bradford area to the south, and forms a link with the even thinner sequence proved in the Winksley Borehole [2507 7151] (Institute of Geological Sciences, 1976) on the Askrigg Block 14 km to the north (Figure 9).

The Marsdenian and Yeadonian strata equate respectively with the 'Middle Grit Group' and the 'Rough Rock Group' of the Leeds and Bradford districts (Edwards et al., 1950; Stephens et al., 1953). Four coarsening-upwards cycles occur within the Marsdenian, each commencing with a marine mudstone and capped by a sandstone. In ascending order, these sandstones are correlated with the East Carlton Grit, Brandon Grit, Guiseley Grit and Huddersfield White Rock of the Leeds–Bradford area and are not given local names. The Yeadonian

**Plate 2**   The Carboniferous–Permian unconformity at Abbey Crags [3550 5583], Knaresborough.

Large-scale, cross-bedded, oolitic limestones of the Lower Magnesian Limestone, Upper Subdivision rest unconformably upon a buried hill of slightly reddened Upper Plompton Grit (L 1814).

strata in the area are represented by marine mudstones, containing the Cancelloceras cancellatum and C. cumbriense marine bands overlain by the Rough Rock Flags and the Rough Rock. The exposures around Farnham are the most northerly of the Rough Rock and Rough Rock Flags, the sedimentology of which, for the area to the south, is described by Bristow (1988). He ascribes the Rough Rock Flags to distal deltaic and delta front environments, and the Rough Rock to a braided river regime on an alluvial plain.

### Details

##### BEDS BETWEEN UPPER PLOMPTON AND EAST CARLTON GRITS

These beds are poorly exposed, and lithological evidence comes from the Farnham Borehole (Figure 8) where the sequence is 18.5 m thick. Here, the seatearth at the top of the Upper Plompton Grit is followed by a 5 mm-thick coal lamina, succeeded by 0.6 m of siltstone and mudstone with plant fragments. This is overlain by mainly marine siltstone and mudstone with the

Reticuloceras gracile Marine Band present about 11.0 m above the base. This marine band yielded the following fauna in the Farnham Borehole: *Sphenothallus* sp., *Lingula mytilloides* (common), *L. squamiformis*, *Orbiculoidea cincta*, *Rugosochonetes* sp., *Euphemites* sp., *Retispira* cf. *undata*, *Anthraconeilo laevirostrum* (common), *A. mansoni*, *Dunbarella* sp., *Pernopecten carboniferus*, *Sanguinolites* sp., *Bilinguites gracilis*, ostracods, fish scales and debris. The top 5.0 m of the unit is siltstone with very fine-grained sandstone laminae; it contains plant debris and is affected by syndepositional slumping.

##### EAST CARLTON GRIT

In the Farnham Borehole this sandstone is 13.89 m thick, dominantly fine grained, and cross-bedded in sets 0.1 to 0.3 m thick. The unit has a gradational base, is extensively bioturbated in places and is capped by a seatearth. At outcrop, the East Carlton Grit is poorly exposed in Bilton Beck [310 583 to 311 584] and on the banks of the River Nidd [313 584; 326 586] where it occurs as limited exposures of fine- to medium-grained flaggy sandstone.

**Plate 3**   Lover's Leap, Plompton Rocks; naturally sculptured crags of Upper Plompton Grit at its type locality of Plompton Park [3552 5363], near Knaresborough (L 1823).

### BEDS BETWEEN EAST CARLTON AND BRANDON GRITS

Like the sequences described above, details of this interval are known only from the Farnham Borehole, where it is 9.7 m thick. The basal 2.15 m comprise a composite mudstone and sandstone seatearth followed by 0.03 m of mainly silty coal. This is overlain by alternating mudstones, siltstones and fine-grained sandstones with abundant *Lingula mytilloides* and sporadic *Phestia sharmani* in the lowest 3.0 m, which constitute a marine band thought to equate with the Bilinguites bilinguis Marine Band (Figure 9).

### BRANDON GRIT

In the Farnham Borehole this sandstone is 10.57 m thick. It has a gradational base and is fine grained, very micaceous and cross-bedded in sets 0.1 to 0.8 m thick; the contact with the overlying mudstone is sharp. At outcrop, the Brandon Grit is massive and was formerly quarried at several localities [3073 5935; 306 593; 317 589]. The topmost 6 m of medium- to fine-grained, cross-bedded sandstone and the overlying siltstones and flaggy sandstones are still exposed in a gully section on Rawson Banks [3204 5880].

### BEDS BETWEEN BRANDON AND GUISELEY GRITS

Known in detail only from the Farnham Borehole, this sequence is 10.05 m thick. The lowest 1.4 m includes marine mudstones which have yielded *Orbiculoidea* sp.. They are overlain by alternations of siltstones and mainly very fine- and fine-grained sandstone in thin and medium beds, many with cross-lamination. A few beds of coarse-grained sandstone are also present.

### GUISELEY GRIT

This unit forms a mappable feature in the ground between Scotton and Scriven, but is not exposed. In the Farnham Borehole, it has a sharp base and comprises 6.94 m of mainly coarse-grained, quartzofeldspathic sandstone. The lowest beds are highly micaceous and the unit is cross-bedded in sets 0.1 to 0.3 m thick.

### BEDS BETWEEN GUISELEY GRIT AND HUDDERSFIELD WHITE ROCK

Details of this 5.6 m-thick interval are known only from the Farnham Borehole. The lowest 1.75 m are a sandstone and silt-stone seatearth, overlain by 0.74 m of mudstone seatearth,

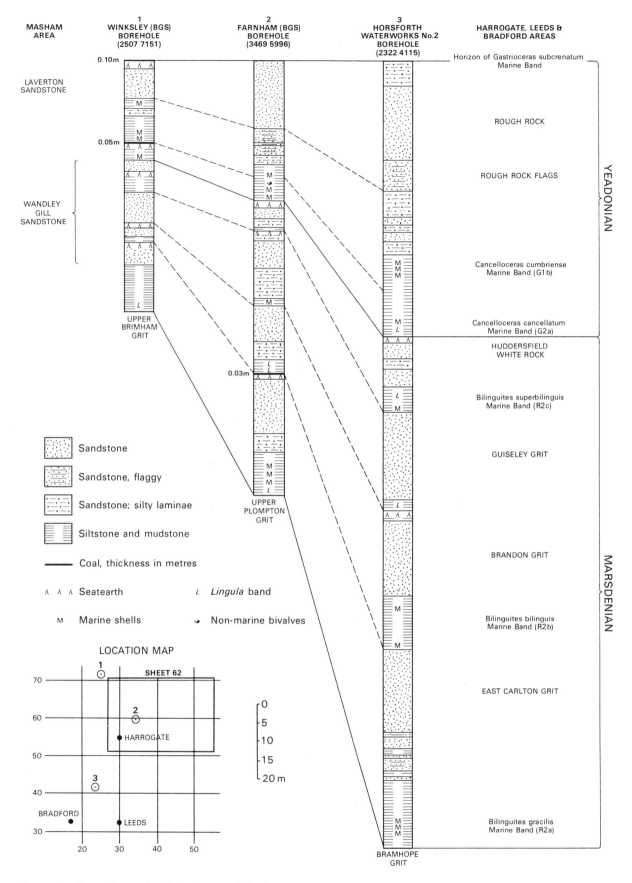

**Figure 9** Late Namurian lithology and thickness variation across the Harrogate district.

**Figure 10** The sequence of Westphalian strata in the east of the district.

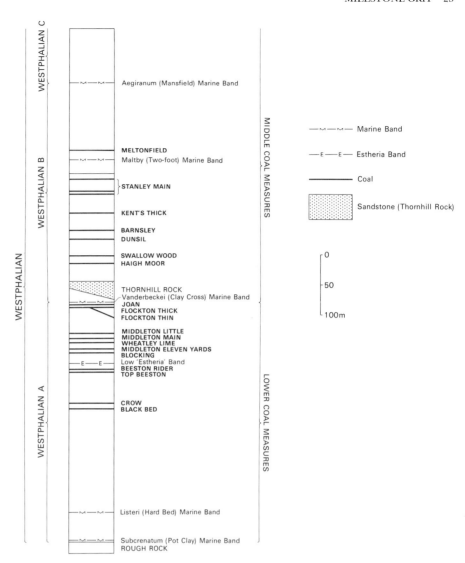

capped by 0.03 m of bright coal. This is followed by 1.06 m of fossiliferous marine mudstone which yielded *Sphenothallus* sp., *Lingula mytilloides* and *Orbiculoidea cincta*. The mudstone passes up through siltstone with sandstone laminae into the gradational base of the Huddersfield White Rock. The marine band is thought to equate with the *Bilinguites superbilinguis* Marine Band (Figure 9).

HUDDERSFIELD WHITE ROCK

This sandstone is not seen at outcrop, but forms a mappable feature between Scotton and Scriven. In the Farnham Borehole, it is represented by only 2.5 m of fine-grained, micaceous sandstone with cross-bedding in thickly laminated and very thin beds. It is capped by a composite mudstone and sandstone seatearth sequence 2.27 m thick.

BEDS BETWEEN HUDDERSFIELD WHITE ROCK AND ROUGH ROCK FLAGS

Proved locally only in the Farnham Borehole, this sequence, 11.75 m thick, is composed dominantly of mudstone. The base is sharp on the underlying seatearth mudstone and the lower beds yielded a fauna including *Lingula mytilloides* (common), *Crurithyris* sp., *Productus carbonarius*, *Microptychis* sp.,

*Aviculopecten* aff. *gentilis*, *Dunbarella* sp., *Phestia stilla*, *Posidonia* sp., *Schizodus* cf. *impressus*, *Agastrioceras carinatum*, *Anthracoceratites* sp. (common), *Cancelloceras* cf. *cancellatum*, *C. crencellatum* (common), ostracods, fish scales and debris; the fauna is indicative of the *Cancelloceras cancellatum* Marine Band. A sideritic limestone, 0.11 m thick, with *Carbonicola deansi* (common), *C. lenicurvata*, *Naiadites* sp. cf. *obliquus* (common), *Geisina arcuata*, fish scales and debris occurs 5.68 m above the base of the unit. This is overlain by more mudstones with a fauna including *Lingula mytilloides*, *Anthracoceratites* sp., *Cancelloceras crenulatum* and *C. cumbriense*, indicative of the *Cancelloceras cumbriense* Marine Band, which is 2.64 m thick. The top of the unit becomes silty, then sandy, passing by gradation into the overlying Rough Rock Flags.

ROUGH ROCK FLAGS AND ROUGH ROCK

In the Farnham Borehole the Rough Rock Flags comprise 6.73 m of thickly laminated and thinly bedded, fine-grained sandstone with siltstone partings. These are overlain by the Rough Rock, which has a sharp transgressive base and is 18.4 m thick. The Rough Rock consists mainly of very coarse- to granule-grained, quartzofeldspathic sandstone with quartz pebbles and mudstone clasts. It has large-scale cross-bedding and decreases in grain size to mainly medium grained in the top 2.0 m.

The Rough Rock sandstone forms a conspicuous feature between Market Flat Gate [343 589] and Lingerfield Terrace [336 598], where about 9 m of coarse quartzofeldspathic sandstone are exposed in several old quarries; quartz pebbles up to 2 cm, and cross-bedded units from 0.3 to 1.0 m thick, are evident. Farther north, the same sandstone forms a small, isolated hill on Low Moor [331 603].

## Coal Measures

Strata of Westphalian A (Langsettian) age occur in small, fault-bounded areas around Farnham [345 605], but are concealed beneath glacial deposits. The sequence, proved in the Farnham Borehole (Figure 8), shows the Rough Rock to be overlain by thinly bedded sandstones and siltstones, all deeply weathered. The basal Westphalian Subcrenatum Marine Band was judged to be missing there, possibly as a result of local contemporaneous erosion (Burgess and Cooper, 1980b). The succeeding strata contain one marine bed, which yielded foraminifera, *Lingula mytilloides*, *Myalina* sp. and *Planolites ophthalmoides*; because of the presence of foraminifera it was taken to be the Listeri Marine Band by Burgess and Cooper (1980b), but could possibly represent the Subcrenatum Marine Band. Farther north, the Spellow Hill Borehole [3516 6234] proved some 52 m of Westphalian sandstones and mudstones, with two coal seams, beneath more than 100 m of Permian strata (Appendix 1).

In the eastern part of the district, extensive coal exploration, including seismic surveys and drilling, was carried out by British Coal between 1978 and 1985 (Figure 6). The North Ouse Coalfield sequence thus proved is comparable in lithology and thickness to that of the Yorkshire Coalfield to the south. The sequence is shown in Figure 10 and comprises about 850 m of Lower and Middle Coal Measures (Westphalian A (Langsettian) and B/C (Duckmantian/Bolsovian)), unconformably overlain by Permian and Triassic strata. Details of the succession and of individual boreholes are confidential. The extent of the coalfield, as shown in Figure 6, is based on Moss (1985); all the boreholes, except Marston Moor which entered Lower Coal Measures in the south-west of the exploration area, are reported to have penetrated potentially workable coals. The Coal Measures dip eastwards and, 6 km east of the district, near Towthorpe, fall to below 1200 m depth and become unworkable. The ground is reported as 'more disturbed underground than at Selby' to the south of the district; this is a result of the faulting present in the Harrogate district (Figure 20). Moss (1985) reported that the majority of the North Ouse coal is clean and suitable for power generation and industrial use.

# THREE

# Permian

For the Harrogate district, William Smith's map of 1821 showed the Yellow or Magnesian Limestone separated by red clay and gypsum from the overlying Knottingley Limestone. These divisions were improved by Sedgwick (1829, pl. 4), who published a detailed description of the Permian strata. Sedgwick divided the sequence into Magnesian Limestone, Lower Red Marl and Gypsum, and Upper Slaty Limestone. This classification was refined by the Geological Survey (Fox-Strangways, 1874; 1908) to Lower Limestone, Middle Marl, Upper Limestone and Upper Marl. This fourfold subdivision of the Permian has since remained almost unchanged and forms the basis for that used on the 1:50 000 Harrogate Sheet. For consistency, therefore, it is also retained in this Memoir

(Figure 11). However, it does not strictly comply with modern formal stratigraphical guidelines (Hedberg, 1976) and has recently been revised (Smith et al., 1986). The proposed new formal stratigraphical names are shown in parentheses (Figure 11 and below), but are not otherwise used here. A review of work on Permian palaeogeography and stratigraphy in the north-east of England has been published by Smith (1989).

Full details of measured sections in the district are given by Cooper (1987); condensed logs of the more important are included in this memoir.

The climax (Asturian phase) of Hercynian earth movements at the end of Carboniferous times caused faulting and folding, and by early Permian times the land surface

**Figure 11** Stratigraphical framework and nomenclature of the Permian strata in the district, and their relationship to coeval strata in the east. Older names, as used in this account, are shown in bold type, formalised names (after Smith et al. 1986) are shown in lighter type. Synonymy is shown by brackets. Thicknesses in metres.

| SYMBOL ON MAP | SEQUENCE AT AND NEAR OUTCROP | | SEQUENCE AT DEPTH AND EAST OF DISTRICT | | ENGLISH ZECHSTEIN CYCLE |
|---|---|---|---|---|---|
| UPM | **UPPER MARL** (Roxby Formation) (10 – 25m) | | **UPPER MARL** (Roxby Formation) (25 – 70m) | **Saliferous Marl** (Roxby Formation) | X |
| | | | | **Top Anhydrite** (Little Beck Formation) | EZ5 |
| | | | | Sleights Siltstone Formation | X |
| | | | | **Upper Halite and Potash** (Sneaton (Halite) Formation) | EZ4 |
| | | | | Upgang Formation | |
| | | | | Carnallitic Marl Formation | X |
| | | | | **Middle Halite and Potash** (Boulby (Halite) Formation) | EZ3 |
| | Evaporites dissolved away at outcrop | | | **Billingham Main Anhydrite** (Billingham Anhydrite Formation) | |
| UML | **UPPER MAGNESIAN LIMESTONE** (Brotherton Formation) (5 – 18m) | | **UPPER MAGNESIAN LIMESTONE** (Brotherton Formation) (18 – 28m) | | |
| MPM | **MIDDLE MARL** (Edlington Formation) (0 – 40m) | | **MIDDLE MARL** (Edlington Formation) (40 – 100m) | Fordon (Evaporite) Formation with anhydrite at base | EZ2 |
| | | | | Kirkham Abbey Formation | |
| | Evaporites dissolved away at outcrop | | | **Hayton Anhydrite** (Hayton (Anhydrite) Formation) | |
| LML | **LOWER MAGNESIAN LIMESTONE** (Cadeby Formation) (0 – 53m) | **UPPER SUBDIVISION** (Sprotbrough Member) | **LOWER MAGNESIAN LIMESTONE** (Cadeby Formation) (50 – 106m) | | EZ1 |
| | | — Hampole Beds — | | | |
| | | **LOWER SUBDIVISION** (Wetherby Member) | | | |
| | **LOWER MARL** (0 – 14m) | | **MARL SLATE** (Marl Slate Formation) (0 – 2m) | | |
| | **BASAL BRECCIA** (0 – 2m) | | **BASAL PERMIAN SANDS** (0 – 46m) | | |
| | Permian strata unconformable on Carboniferous strata | | | | |

**Plate 4** The Carboniferous–Permian unconformity at Newsome Bridge Quarry [3789 5143], near North Deighton.

A buried hill of slightly reddened Upper Plompton Grit is overlain by the Lower Magnesian Limestone, Lower Subdivision. At the base of the limestone, a breccia of Carboniferous sandstone clasts occurs on the flank of the hill. The lower part of the limestone comprises a patch reef with an uneven upper surface overlain by evenly bedded oolitic and pisoidal grainstone (A 5225).

took the form of a major land-locked basin extending from eastern England across to Germany and Poland (Smith, 1970b, 1980; Ziegler, 1982; Glennie, 1983). The Harrogate district lay near the western margin of this basin, in tropical palaeolatitudes.

Newly uplifted areas were subjected to intense, mainly subaerial erosion in a desert environment, which sculpted the upstanding features, especially around Harrogate and Knaresborough, into steep-sided hills with a relief of up to 30 m. These hills, subsequently buried by the carbonate deposits of the Lower Magnesian Limestone, are exposed in Knaresborough Gorge (Fox-Strangways, 1908; Burgess and Cooper, 1980a) and elsewhere. Here Carboniferous sandstones below the unconformity commonly show signs of leaching and red-brown and purple staining associated with the subaerial erosion (Plate 2). At Newsome Bridge Quarry [379 515] local pockets of breccia are present above the unconformity (Plate 4). Dune sands that were transported across the land surface have also been preserved at the base of the Permian sequence, mainly in the east of the district (Figure 12).

During the late Permian, the depositional basin underwent a series of cycles of flooding and evaporation, possibly caused by glacioeustatic changes: these are the five English Zechstein Cycles EZ1–EZ5 (Smith, 1970b and Figure 11). Where fully developed, individual cycles show an upward progression in evaporitic sedimentation from: carbonate to anhydrite to halite to sylvinite, reflecting the crystallisation of increasingly soluble minerals. The carbonate and sulphate phases are best developed near the margins of the basin and those of the EZ1 and EZ3 cycles, the Lower and Upper Magnesian Limestones, and the Hayton and Billingham Main Anhydrites, are generally well preserved in the Harrogate district (Figure 11). Halite and sylvinite deposits, however,

are restricted to the more central parts of the basin, near the Yorkshire coast and farther east (Smith, 1974a); westwards, their lateral equivalents at outcrop are the red-brown terrigenous and marginal marine mudstones of the Middle Marl (Edlington Formation; EZ1–2) and the Upper Marl (Roxby Formation; EZ3–5). The boundaries of the cycles are not necessarily coincident with the formational boundaries (Figure 11).

## DESCRIPTION OF SEQUENCE

### Basal Breccia

At Newsome Bridge Quarry [379 515], on the flank of a buried hill of Carboniferous sandstone, a 1 m sequence of dolomite-cemented breccia is present. The breccia clasts are slabs of the local sandstone, up to 0.9 m across, arranged in rough imbricate piles cemented with sandy dolomite (Smith, in press). Conglomerate beds were also recorded by Tute (1892) from the base of the Lower Magnesian Limestone at Markington (exact site unknown, perhaps near [2900 6522]). Here Tute recorded a bed about one yard (0.92 m) thick exposed with a few yards lateral extent in Markington Beck. This bed is recorded as consisting of waterworn pebbles of Carboniferous limestone and sandstone, with angular fragments of Magnesian Limestone, embedded in a matrix of harder Magnesian Limestone. In the Spellow Hill Borehole (Appendix 1), the base of the Permian sequence is marked by a 2.44 m-thick conglomerate bed, with its base at a depth of 121.62 m, resting on reddened Carboniferous shales with ironstone nodules. These beds may equate with or be older than the Basal Permian Sands present at depth in the east (see below and Figure 12); because of their limited extent at outcrop, the brec-

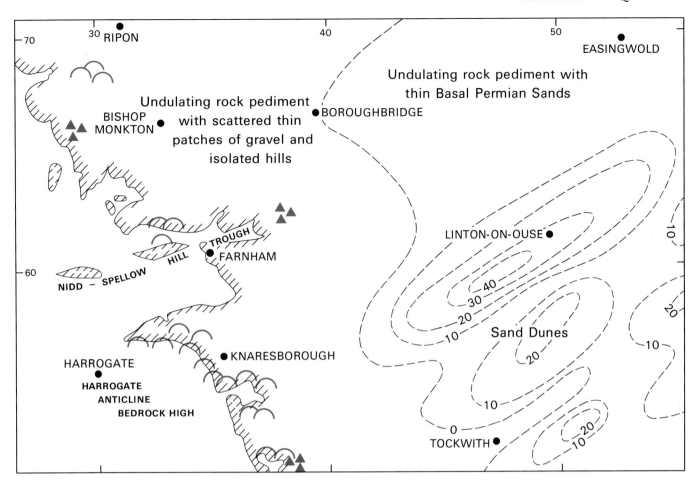

—10— Approximate isopachytes for the Basal Permian Sands;
contour interval 10 metres

ⵍⵍⵍⵍ Western limit of Lower Magnesian Limestone

▲▲ Basal Breccia

⌒⌒ Hills of Carboniferous bedrock (mainly buried beneath
Lower Magnesian Limestone)

Scale 0   1   2   3   4   5 km

**Figure 12**   Generalised isopachyte map for the Basal Permian Sands, also showing the occurrence of the Basal Breccia and the location of buried hills of Carboniferous strata beneath the Lower Magnesian Limestone.

cias are not differentiated from the Lower Magnesian Limestone on the map.

**Basal Permian Sands**

The Coal Measures strata, preserved at depth in the Vale of York, were worn down to a rock pediment or stony desert. Within this intracontinental desert, sand dunes developed; some of these are preserved buried at the base of the Permian sequence in east and the south-east of the Harrogate district (Figure 12). Penetrated by numerous exploration boreholes, the Basal Permian Sands are seen to be lenticular, ranging from a metre or two to

a maximum thickness of 46 m. This thickness variation is comparable to that recorded for the Basal Permian Sands in Durham where both Smith and Francis (1967) and Steele (1983) described south-west- to north-east-trending draa. The Vale of York sand ridges have a similar south-west to north-east orientation (Figure 12), a trend which is perpendicular to that suggested by Versey (1925) for the district immediately to the south. In boreholes, the sands are bluish grey and their aeolian origin is attested by their typically rounded and frosted sand grains. These sands are comparable with the Durham Yellow Sands (Smith and Francis, 1967; Krinsley and Smith, 1981), the Basal Permian Sands of south Yorkshire and Nottinghamshire (Edwards et al., 1950; Mitchell et al., 1947; Versey, 1925)and the Rotliegendes of the North Sea (Glennie, 1982, 1983). The Basal Permian Sands at outcrop in Yorkshire and Durham weather to yellow and orange-yellow due to the oxidation of ferruginous pellicles on their grains.

## Lower Marl and Marl Slate (Marl Slate Formation; EZ1)

As the Zechstein basin rapidly flooded, the Basal Permian Sands were quickly submerged by the Zechstein sea (Smith, 1970). In this were deposited the Lower Marl and Marl Slate (Marl Slate Formation), an evenly laminated, carbonaceous, dolomitic and calcareous siltstone sequence, up to a metre or so thick. The Marl Slate is recorded as representing a period of condensed deposition in partly anoxic conditions (Deans, 1950), possibly with a freshwater influence (Turner and Magaritz, 1986). It is present locally in the east of the district, where it has been proved in exploration boreholes; it has not been observed at or near outcrop. However, the Lower Marl, a thin sequence of grey calcareous mudstones, in excess of 3.5 m thick, is locally present at the base of the Lower Magnesian Limestone in a cliff section [2926 5978] 150 m south of Rock Farm, Nidd. This appears to be a deposit local to the Nidd–Spellow Hill trough (Figures 12 and 13), a feature apparently located along the line of the Ripley–Staveley Fault Belt. It is proved again in the Spellow Hill Borehole (Appendix 1) where equivalent deposits overlie the Basal Breccia and comprise 'blue shales' 13.41 m thick. Because of their limited extent at outcrop, these deposits are not differentiated from the Lower Magnesian Limestone on the map.

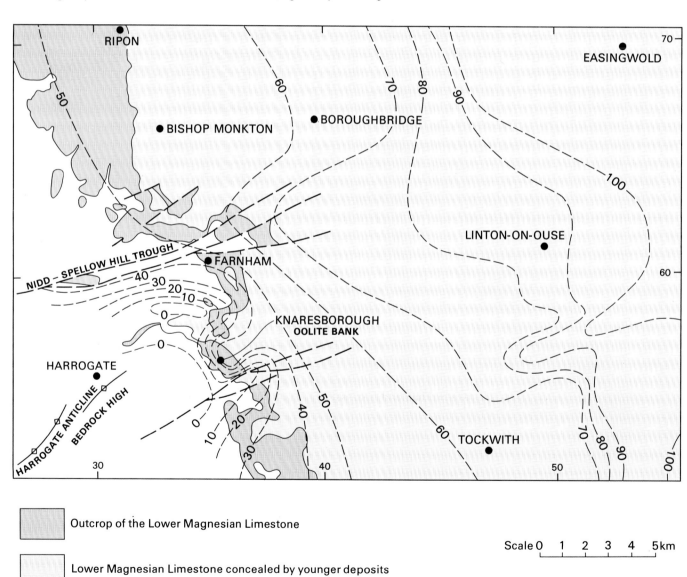

Outcrop of the Lower Magnesian Limestone

Lower Magnesian Limestone concealed by younger deposits

—10— Approximate isopachytes for the Lower Magnesian Limestone; contour interval 10 metres

Scale 0  1  2  3  4  5km

**Figure 13** Generalised isopachyte map for the Lower Magnesian Limestone in the Harrogate district.

## Lower Magnesian Limestone (Cadeby Formation; EZ1)

The Lower Magnesian Limestone (Cadeby Formation) represents the carbonate phase of the first (EZ1) cycle in the Zechstein basin. Because the Harrogate district was marginal to this basin, sedimentation was laterally variable, as deposition gradually overwhelmed the pediment and marginal hill topography. The formation largely comprises dolomitic limestone and ranges in thickness from 0 to 103 m, generally thickening eastwards (Figure 13). Two subdivisions of the limestone, Lower and Upper, were originally recognised by Edwards et al. (1950) in the district to the south. Smith (1968) described the Hampole Beds separating the two subdivisions and subsequently (1974a and b) traced these units northwards through the Harrogate district. Smith et al. (1986) formalised the nomenclature of these subdivisions and introduced the names Wetherby Member for the Lower Subdivision, and Sprotbrough Member for the Upper Subdivision (Figure 11); the Hampole Beds are split between the two members. These divisions do not form mappable units, but are useful for describing the sequence and sections.

The Lower Magnesian Limestone was deposited as a carbonate wedge thickening eastwards to a marginal reef belt about 30 to 40 km away (Smith, 1974a, 1989; Taylor and Colter, 1975). The Lower Subdivision represents deposition in a shallow-water, semirestricted inner shelf or lagoon on a carbonate platform, with sporadic patch reefs developed on minor bedrock highs. The overlying Hampole Beds were formed during an episode of marine regression and transgression which left the marginal parts of the carbonate platform exposed to subaerial influences for a time (Smith 1968, 1974a and b); these beds are well seen at outcrop, but become less well developed eastwards and northwards. The Upper Subdivision at outcrop is largely characterised by an oolitic facies with either large-scale sand waves typical of oolite bank deposits or evenly bedded deposits suggestive of a lagoonal environment.

The Lower Subdivision (Wetherby Member) of the Lower Magnesian Limestone varies in thickness near outcrop from 0 to 40 m. The earliest sediments are interbedded calcareous mudstones and thin-bedded dolomites; like the Lower Marl these were restricted to the low-lying areas such as the trough through Nidd [294 599] and Spellow Hill [381 623], which was probably fault controlled. There followed a sequence of evenly bedded and cross-bedded dolomites with minor channel deposits, formed throughout the district, except on the upstanding areas around Harrogate and Knaresborough where the Lower Magnesian Limestone is thin (Figures 13 and 14). These dolomites contain abundant shelly and bryozoan debris, some oolitic material and in a few places, as at South Stainley [3145 6276] and near Brearton [3205 6195] (Smith, 1974b), small bryozoan patch reefs typified by the colonial bryozoan *Acanthocladia*; these reefs are similar to those described by Smith (1981) from farther south in Yorkshire. An inferred patch reef (Smith 1974b and in press) is also present at Newsome Bridge Quarry [3790 5144] where an unbed-

ded, elongate mass of recrystallised dolomitic limestone caps a hill of Carboniferous sandstone. The limestone mass has a sharp contact with, and is surrounded by, bedded peloidal grainstones which have yielded abundant bivalves (mainly *Bakevellia binneyi*) and some bryozoa suggestive of a reef origin. In the district, the Lower Subdivision is interpreted as a shallow-water carbonate deposit formed in a semirestricted inner shelf or lagoon.

The top of the Lower Subdivision (Wetherby Member) and the lower part of the overlying Upper Subdivision (Sprotbrough Member) are formed by the Hampole Beds (Smith, 1968) which are split between the two units (Smith et al., 1986). They comprise a thin impersistent sequence of fenestral (cellular) dolomites and mudstones, 0.3 to 0.6 m thick, typical of an intertidal to supratidal environment. These beds record a marine regression event, with the formation of a subaerially exposed surface upon which the fenestral dolomites developed, followed by a mudstone and thin dolomite sequence developed during the subsequent transgression. These beds are recognisable at Monkton Moor Quarry [307 653] and Rock Cottage Quarry [300 650] (Smith, 1968), both near Wormald Green; they may also equate with a sedimentary break in the sequence at Lime Kiln Farm quarries [3236 6293]. The Hampole Beds are well developed in a belt which represents shallow-water to subaerial conditions of sedimentation, and largely coincides with the Lower Magnesian Limestone outcrop. Down dip, and to the north in the Thirsk district, where a deeper water facies is present, the beds show a less marked sedimentary change and faunal hiatus, such as that noted by Pattison (1978) in the Aiskew Bank Farm Borehole. The Hampole Beds have not been recognised in the exploration boreholes in the east of the district.

The Upper Subdivision of the Lower Magnesian Limestone is present at outcrop in two main facies. The northern facies, extending from Wormald Green [300 650] northwards, generally consists of evenly bedded, commonly oolitic dolomite with traces of cross-lamination, in sets 0.1 to 0.5 m thick; shell debris, especially of *Bakevellia* and *Schizodus*, is common, and trace fossils (Tute, 1890) also occur. This facies appears to be a lagoonal sequence developed behind a barrier, probably an oolite bank. At Quarry Moor [308 691], the upper part of the subdivision is formed by an 11 m sequence of laminated algal stromatolites with interbedded oolites and mudstones. Tepee-like structures suggestive of displacive evaporite formation occur, together with probable dissolution residues after gypsum (Smith, 1976 and Plate 6). These beds suggest a shallow-water to emergent sabkha facies and mark a transition into the completely evaporitic regime of the ?Hayton Anhydrite, at the base of the succeeding Middle Marl.

The southern facies of the Upper Subdivision is exposed from Lime Kiln Farm [3236 6293] southwards. It is mainly oolitic, with large-scale cross-bedding, such as that displayed along the Knaresborough gorge (Plate 5) where cross-bedded sets up to 18 m thick occur around Bunkers Hill Quarry [3517 5657]. The cross-bedding has a subaqueous dune or sandwave origin (similar to the

**Plate 5** Lower Magnesian Limestone, Upper Subdivision at the House in the Rock [3513 5645], Knaresborough. The limestone is oolitic and forms large-scale (5 to 18 m high) cross-bedded units interpreted as submarine sand waves (L 1808).

present day ooid shoals of the Bahama Banks), and indicates that depositional currents in the Harrogate district came mainly from the east (Kaldi, 1980a and Figure 14). Around Knaresborough the upper surface of a major dune bank has been preserved, so that the junction with the overlying Middle Marl is undulating. Knaresborough stands on an exhumed dune crest and the Lower Magnesian Limestone, beneath the highest part of the town, is about 40 m thick. The eastward fall in slope away from Knaresborough centre is attributable mainly to tectonic dip, whereas that to the west, north and south results from a rapid sedimentary thinning of the Lower Magnesian Limestone along the face of the dune (Figure 13). The westwards thinning is accentuated by overlap of the Upper Subdivision onto the sub-Permian land surface (Figure 14), so that the Lower Magnesian Limestone has a total thickness of only 7 m near Dropping Well Farm [3459 5707]. Northwards from here it thins to only 0.5 m at Foolish Wood [3413 5722], thickening again to 7.35 m at Scotton Banks [3313 5807] (Burgess and Cooper, 1980a). Southwards, the Lower Magnesian Limestone thins to about 10 m in the vicinity of Goldsborough Mill [3676 5590], where the basal unconformity is exposed and where the Middle Marl is seen a little way upstream [3675 5603].

As its name implies, the Lower Magnesian Limestone is largely composed of dolomite and calcitic dolomite. This is the result of complex diagenetic processes, for details of which reference should be made to Clark (1980), Fuzesy (1980), Harwood (1980), Kaldi (1980a,b) and Kaldi and Gidman (1982). The initial dolomitisation and subsequent alteration of the rock have generally destroyed some or all of the small-scale primary sedimentary structures, reducing the rocks to a porous crystalline mesh with numerous vughs. Alteration was patchy, commonly forming amorphous masses within otherwise well-bedded sequences. This absence of any obvious pattern

of alteration makes the prediction of rock quality for quarrying very difficult.

### Details

*Ripon to Wormald Green*

Near Fountains Abbey, the basal Lower Magnesian Limestone unconformity crosses the River Skell near Half Moon Pond [280 686]. About 1.2 km east of here, the unconformity is again recorded at Mackershaw Trough [2924 6859], where numerous small exposures prove an 8 m-high buried hill of fine-grained Carboniferous sandstone covered by poorly exposed buff-coloured dolomite. Taking into account the eastward dip of the sequence, the true relief at the unconformity is probably in the order of 30 m.

On the east side of Mackershaw Trough [2925 6970], a disused quarry exposes about 10 m of mainly buff-coloured, soft, compact, and minutely vesicular dolomite. Bedding is largely obscure, except for two prominent beds 2 m and 4 m above the base of the sequence; the lower one is capped by a 0.01 m clay bed.

On the left bank of the River Skell near Hell Wath [2996 7010] 2.5 m of buff, thin-bedded, soft dolomite with thin hard beds containing vughs is present. The hard beds contain sporadic fossils including a probable vermiform foraminiferan (*Orthovertella?*) and shells of *Bakevellia binneyi, Cardiomorpha?* and *Schizodus obscurus* with sporadic thin-walled curved tubes which are probably productoid spines.

At Quarry Moor [308 691], the upper 13 m of the Lower Magnesian Limestone, almost to its contact with the overlying Middle Marl, is exposed. This dominantly dolomite and calcitic dolomite sequence is fully described and interpreted by Smith (1976). The lowest unit, 3.7 m thick, is fairly massive, with cross-lamination in the middle. This is overlain by about 3.5 m of slightly oolitic, algal-laminated dolomite, followed by 2.25 m of generally thinly bedded and algal-laminated dolomite. The upper 3 to 4 m of the sequence comprises laminated and thinly bedded dolomite, calcitic dolomite, dolomitic limestone and limestone (Plate 6). These are interbedded with thin earthy clay layers and beds of dolomitic mudstone, some of which are

**Figure 14** The geology of Knaresborough gorge showing the Knaresborough Lower Magnesian Limestone oolite bank thinning to the west (it also thins to the north and south; Figure 13). The Lower Magnesian Limestone is overlapped by the Middle Marl and the Upper Magnesian Limestone, which rest unconformably on the upstanding Carboniferous strata of the Harrogate Anticline.

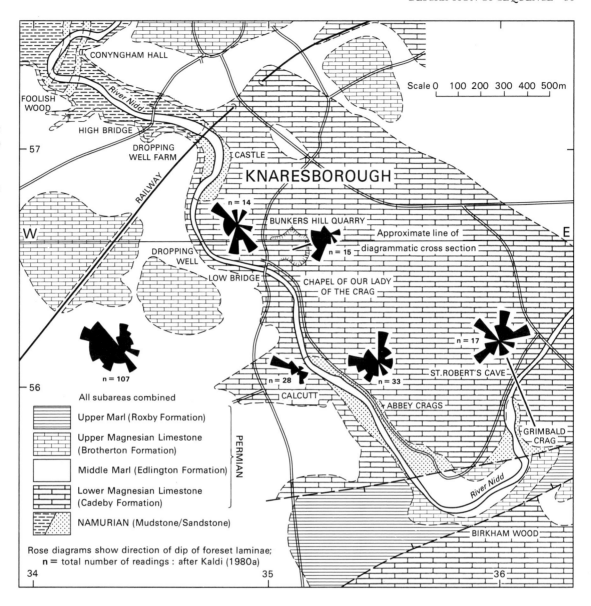

All subareas combined

Upper Marl (Roxby Formation)

Upper Magnesian Limestone (Brotherton Formation)

Middle Marl (Edlington Formation)

Lower Magnesian Limestone (Cadeby Formation)

PERMIAN

NAMURIAN (Mudstone/Sandstone)

Rose diagrams show direction of dip of foreset laminae;
**n** = total number of readings : after Kaldi (1980a)

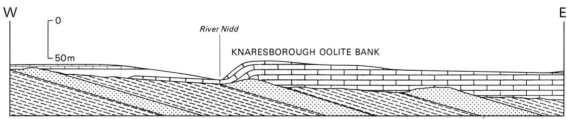

DIAGRAMMATIC CROSS-SECTION ILLUSTRATING RELATIONSHIPS BETWEEN CARBONIFEROUS AND PERMIAN STRATA

residues after the dissolution of evaporites. In places, brecciation and tepee-like structures in these upper beds suggest expansion of the beds during lithification at depths just below the surface of deposition.

Beds similar to those at Quarry Moor were formerly seen in the disused railway cutting north-east of Monkton Moor Quarries [3100 6570]. Here, approximately 7 m of buff-coloured, thinly laminated dolomite, with probable algal stromatolite structures, were present. A prominent 0.01 m-thick brown mudstone bed occurred about 2 m above the base of the expo-

sure. Lithologically, the dolomite varied from hard and compact to coarsely crystalline.

For the northern of the Monkton Moor Quarries [307 653], Dr D B Smith furnished details (in Cooper, 1987) from which the following description is summarised. The quarry exposes a maximum of 13.8 m of mainly dolomite, with subordinate mudstone beds. Part of the Lower Subdivision (Wetherby Member), the Hampole Beds and all of the Upper Subdivision (Sprotbrough Member) are exposed. The basal 6 m of the sequence comprise buff-coloured dolomite in thin, medium and

**Plate 6**  Lower Magnesian Limestone, Upper Subdivision, at Quarry Moor [3080 6921], near
Ripon. This is the topmost part of the Lower Magnesian Limestone, which shows contortion
and brecciation of the beds due to evaporite growth, flow and dissolution (Smith, 1976) (L 1805).

thick beds with ripple- and cross-lamination, channel and fill
structures. Lithologically, this part of the sequence is fine-
grained and oolitic, with some shell debris and stromatolite
flakes. It is overlain by the Hampole Beds, the lowest bed of
which comprises 0.1 to 0.2 m of hard, fine-grained dolomite
with fenestrae and a dense crust-like top. This bed also has
traces of ooliths and cross-bedding. Its upper surface is a discontinuity and represents alteration of the pre-existing sediment by subaerial exposure. Above this, there is 0.05 to 0.13 m
of brown and olive-coloured clay, with a 0.02 m dolomite bed
in the middle. These beds are capped by 1.1 m of buff
dolomite with brown clay laminae; the unit is thinly bedded,
finely oolitic, and has planar and ripple-laminated bedding; it
marks the base of the Upper Subdivision. It is overlain by 1.2 m
of oolitic, cross-laminated dolomite in medium and thick beds,
capped by a discontinuous clay bed (0 to 0.02 m). The top
3.68 m exposed comprise oolitic dolomite, in medium and
thick beds, with gentle wedge-bedding and large-scale cross-
lamination. The top of this unit is uneven, the uppermost 0.05
to 0.08 m containing irregular nodules of blue-grey limestone,
above which there is 0.08 m of buff and grey dolomitic mudstone and clay. This is locally overlain by red mudstone forming the basal 0.5 mm of the Middle Marl hereabouts.

The nearby quarry at Rock Cottage [300 650] shows approximately 24 m of the succession, the rock having been formerly
extracted for agricultural limestone and inferior roadstone.
The lowest part of the section comprises 11.03 m of slightly
oolitic dolomite and dolomitic limestone, in mainly thin and
medium beds, with a prominent 0.7 m bed of oolite 1.6 m
below the top. Many of the bedding planes are stylolitic, and
vughs filled or lined with calcite are common. At a level 6 to
7 m above the base of the section, numerous siliceous nodules
occur; similar nodules were recorded from here by Tute
(1894). These beds also yielded one sample with angular
monoclinic cavities, up to 0.03 m across and ornamented with
fine bars parallel to their edges; these cavities were possibly left
after the dissolution of gypsum crystals. Also at this level, sporadic horizontal sinuous trace fossils, oval in section and with a
diameter of about 0.01 m, are present; similar trace fossils were
recorded from this locality by Tute (1890). The lowest 10 m of
the sequence here yielded abundant fossils including: *Agatham-
mina aff. milioloides, A. pusilla, Cyclogyra* sp., ?productoid spines,
*Dentalium sorbii*, turreted gastropods, *Bakevellia binneyi, ?Permopho-
rus costatus, Phestia speluncaria*, and *Schizodus obscurus*. The Hampole Beds are represented by an impersistent (0 to 0.1 m) bed
of hard, buff-grey, porcellanous dolomite overlain by up to 0.03

m of brownish grey clay; in older parts of the quarry, faint fenestrae are present in the dolomite bed. The uppermost 8 to 9 m of the succession, representing the Upper Subdivision, comprises mainly oolitic and shelly dolomite in medium and thick beds with sporadic traces of cross-bedding. Impersistent mudstone partings are common, as are stylolitic contacts and numerous vughs.

On the southern limit of Monkton Moor Quarries, a water borehole [3079 6515], with a surface elevation of about 84 m OD, proved the Lower Magnesian Limestone to a depth of 46.3 m, resting on grey grit to 50.6 m, blue shale to 51.8 m and red grit to the bottom of the hole at 61.9 m. The 'grey grit' may in fact be coarse dolomite, the 'blue marl' Lower Marl and the 'red grit' the only true Carboniferous rock in the borehole. The borehole is located near the top of the limestone formation, which is probably at least 48 m thick hereabout; but it may, however, be as much as 52 m thick if the alternative interpretion of the hole is accepted.

### Burton Leonard to Farnham

Yew Bank Quarry [3074 6404], 800 m north of South Stainley, exposes approximately 11 m of beds. The lower 6 m comprise dolomite and calcitic dolomite in thin and medium beds with partings (0.1 to 0.3 m) of dolomitic mudstone. The upper 5 m are soft dolomite in medium and thick beds (up to 0.4 m).

In the vicinity of Burton Leonard, exposure is poor because of extensive drift cover, but the Lower Magnesian Limestone is proved in two water boreholes. The first [3206 6378], about 600 m west of the village, with a surface elevation of 62.5 m above OD, penetrated drift to 3.7 m, Middle Marl to 10.7 m, Lower Magnesian Limestone to 63.7 m, on 'grit' to the bottom of the hole at 71.9 m; the thickness of the limestone proved here is 53 m. East of Burton Leonard another borehole [3350 6404], with a surface elevation of 61.6 m above OD, proved drift to 2.1 m, Middle Marl to 7.9 m, Lower Magnesian Limestone to 50 m, 'blue gritty hard rock' to 57.9 m, 'coarse gritty rock' to 61.6m and hard white limestone to the bottom of the hole at 62.2 m. The presence of white limestone at the bottom of the hole suggests that this may still be part of the Lower Magnesian Limestone, in which case the 'gritty rocks' may possibly be coarsely crystalline dolomitic limestone. This borehole suggests a minimum thickness for the Lower Magnesian Limestone of 42.1 m and a maximum of 54.3 m.

Near Burton Leonard, two quarries are present 100 m [3236 6293] and 300 m [3238 6306] north of Lime Kilns Farm. Both show similar sequences, each including parts of the Lower and Upper subdivisions, and the Hampole Discontinuity; however, the typical Hampole Beds are not developed. About 11 m of the Lower Subdivision are exposed. The beds are mainly medium, thick and very thick bedded with faint parallel banding and cross-bedding on a 0.1 to 0.5 m scale. A few split into thin beds and some are laterally impersistent. The rock is all dolomite, porous and crystalline, with relict ooliths and sporadic pisoliths in the upper part. At the top of this unit there is an uneven erosion surface the Hampole Discontinuity, which has a relief of about 1 m and which cuts into the beds below. On this surface there is an irregular bed of soft silty dolomite 0.01 to 0.03 m thick, possibly equating in part with the Hampole Beds. The highest parts of the quarries expose 3 to 3.5 m of the Upper Subdivision, developed in the sand-wave facies that typically predominates southwards from here. The beds are generally very thinly to medium bedded, but have cross-stratification on a 2 to 3 m scale. The rock is white to buff and crystalline; no ooliths were observed.

In the vicinity of Stainley Hall there are numerous small quarries. One [3132 6258] exposes about 5 m of buff partly crystalline dolomite, in thin to medium beds, with scattered shells and plant debris. Nearby, another [3142 6275] shows 1.25 m of dolomite, the lower 0.15 m being laminated and fine-grained, with bryozoa and shells, the upper 1.1 m being massive and crystalline, with vesicles and bryozoa. A small exposure at the northeast end of the quarry [3145 6276] shows only 0.3 m of beds, comprising a pocket of shelly material with a rounded upper surface bound by a 0.01 m-thick skin of fine-grained, laminated dolomite with dark carbonaceous layers. This is possibly a small patch reef overlain by an algal stromatolite skin. Fossils present in the shelly material include: *Acanthocladia anceps, Thamniscus* sp., *Bakevellia binneyi, Liebea squamosa, Parallelodon striatus, Pseudomonotis speluncaria* and *Schizodus obscurus*.

In a disused quarry at Limekiln Hill [3205 6195], 350 m west of Hill Top Farm, Brearton, a section approximately 30 m wide and 4 m thick is exposed. It is composed of irregularly shaped masses of buff-coloured dolomite, up to 3 m wide and 1.5 m high, enclosed within an irregularly bedded sequence of similar rock. Bryozoan fragments are abundant throughout the sequence and the irregularly shaped masses have the form of patch reefs (Smith, 1974b), similar to those recorded by Smith (1981) farther south.

From Limekiln Hill the Lower Magnesian Limestone escarpment strikes eastwards for a distance of 700 m to Simon Slack Wood [3274 6186], then a further 2.3 km to Occaney. South of Simon Slack Wood, sandstone fragments appear in the soil and there are two small Carboniferous sandstone quarries; one of these [3270 6162] exposes 2.6 m of fine-grained, cross-bedded sandstone in beds 0.2 to 1.0 m thick. Fifty metres north-east of here, dolomite fragments abound in the field brash; mapping suggests that the Lower Magnesian Limestone unconformity rises over a buried hill of sandstone at this point.

At Occaney [353 618] the strike of the Lower Magnesian Limestone swings sharply southwards and the outcrop pattern is disrupted by the Ripley–Staveley Fault-Belt (Figures 13 and 20). The Limestone is stepped eastwards in two fault blocks. In the more southerly of these, it is exposed as field brash just west of Loftus Hill [3697 6147]. East of Loftus Hill, a borehole [3718 6154], surface level 61 m above OD, proved clay to 3.7 m, and limestone to 23.8 m, on red grit. The Spellow Hill Borehole (Appendix 1), 1.3 km north-east of here, proved the full thickness of the limestone, but the log is open to several interpretations. The basal conglomerate and blue shale of the Lower Marl have already been discussed (see above). The base of the Lower Magnesian Limestone is fairly certain at a depth of 105.77 m, but the depth to the top of the limestone is debatable. It could be the base of the Middle Marl grey shale at a depth of 34.98 m. However, if the 'Limestone Marl' is in fact gypsum, then the top of the limestone would be at 53.34 m. Depending on which interpretation is taken, the limestone has a thickness of 70.79 m or 52.43 m.

From Loftus Hill, an east–west-trending fault steps the Lower Magnesian Limestone outcrop 2 km westwards again to Farnham. Here, it forms a marked escarpment running south-south-east for 2 km to Butterhills. North of Farnham, a small quarry [3523 6073] exposes 3 m of buff-coloured oolitic dolomite with abundant shells. The sequence is irregularly bedded, with red-brown mudstone partings. Fossils present here include: *Acanthocladia anceps*, indeterminate gastropods, *Bakevellia binneyi, Permophorus costatus, Schizodus obscurus* and ostracods.

Another quarry [3543 6029], south-east of Farnham, exposes up to 8 m of orange-brown dolomite. The rock is minutely vesicular and variable from soft and powdery to hard and crystalline. The beds are thin and medium, but form massive cross-bedded sets up to the height of the quarry; these structures are typical of the sand-wave facies that characterises the Upper

Subdivision (Wetherby Member) of the Lower Magnesian Limestone in its southern development.

At Butterhills, the limestone was seen in a disused railway cutting [3639 5890] which was in the process of being filled with domestic refuse. Here 9.7 m of buff and yellow dolomite are present. The rock is fine to coarse grained, with relict ooliths and dolomite-lined vughs. The beds are mainly medium and thick, with a few very thick units; cross-bedding in sets up to several metres high is evident.

At the southern end of the Farnham escarpment, the Lower Magnesian Limestone thins towards the variable sequence seen at Knaresborough. It is also folded over the Bilton–Scriven Monocline (Figure 20) so that the outcrop swings westwards about a kilometre to Scriven. Here, at Coney Garth [3505 5892], approximately 6 m of beds near the base of the Lower Magnesian Limestone are exposed. The lowest 4 m or so comprise buff and yellow oolitic and pisolitic dolomite with shell fragments. A smooth cylindrical burrow, 0.05 m in diameter, occurs on a bedding plane near the bottom of the section. The beds range from thin to thick and massive, with traces of cross-bedding, and include one well-developed lenticular channel near the middle of the unit. Above this dolomite unit there is a lenticular bed of red-brown calcareous mudstone with dolomite laminae and traces of ripple-drift lamination. The top of the section comprises 2.1 m of buff-coloured, fine-grained dolomite forming a massive unit with calcite-lined vughs and traces of thick bedding.

### Nidd

Downthrown between faults of the Ripley–Staveley Fault Belt (Figure 20), the Lower Magnesian Limestone is preserved as an outlier at Nidd. Several exposures combine to give about 32.5 m of the sequence. The lowest part is exposed in a cliff section [2926 5978] by the River Nidd south of Rock Farm. Here, the Lower Marl (see above), comprising 3.5 m of grey and yellow, silty, dolomitic mudstone with thin beds of dolomitic limestone, occurs at the base of the section; these beds may be similar to the 'blue shale' recorded below the Lower Magnesian Limestone in the Spellow Hill Borehole (Appendix 1). Above this are 13.8 m of rubbly and nodular dolomite with vughs, capped by a 0.4 m bed of stromatolitic dolomite. This is overlain by 2 m of fine-grained dolomite and then by approximately 5 m of dolomite, which is coarsely oolitic and pisolitic, with grapestones and oncolite-coated pebbles. This distinctive oncolitic unit can be traced to the face on the south side of a nearby quarry [2936 5984], where approximately the upper 3 m of it are exposed. Above this, a 0.1 m bed of dolomite with stromatolitic laminae is followed by another oolitic and oncolitic dolomite bed about 3 m thick. This last bed can be correlated with the lowest part of the sequence seen in the eastern end of the quarry [2941 5988], where the upper 2.7 m of it are seen. This unit here comprises a soft yellow dolomite which becomes harder towards the base; it is oolitic, with pisoliths and flat oncolite-coated pebbles up to 0.05 m thick and 0.1 m long. Above, four thin beds of alternating hard and soft dolomite form the next 0.2 m of the sequence, overlain by 0.2 m of fine-grained dolomite with vughs and possible stromatolitic laminae. The top of the sequence comprises approximately 4.5 m of yellow fine-grained dolomite, possibly oolitic, occurring in laminated and thin beds with possible low-angle cross-bedding.

### Knaresborough

In the Knaresborough area, under the town, the Lower Magnesian Limestone forms a thick oolite bank which thins to the north, west and south (Figures 13 and 14). The attenuated sequence to the north is exposed in the banks of the River Nidd at Scotton Banks [3313 5807], where it is 7.35 m thick and has a sharp unconformable base on coarse-grained Carboniferous sandstone. The lowest 2.25 m of limestone and dolomite is horizontally laminated and sandy, with pebbles, shells and sporadic vughs. This is followed by 4.9 m of oolitic dolomite with vughs in medium beds, capped by 0.2 m of hard dense dolomite with horizontal laminae. The dolomite sequence is overlain by the red-brown mudstones of the Middle Marl, here about 15 m thick.

At Conyngham Hall Farm [3425 5742] the sequence is further attenuated, with only 1.2 m of the Lower Magnesian Limestone present. The Carboniferous siltstone beneath the unconformity is coloured purple. The succeeding 1.2 m of dolomite is sandy and contains vughs and, at its base, fragments of Carboniferous sandstone. This is followed by the Middle Marl which comprises 0.65 m of fine-grained, micaceous, red-brown sandstone overlain by 2.0 m of red, yellow and purple mudstone.

Between Foolish Wood [3413 5722] and Dropping Well Farm [3459 5707], the Lower Magnesian Limestone is intermittently exposed in the cliff on the south-west side of the River Nidd. It is a thin (0.5 to 7.0 m) sandy dolomite, overlain by red mudstones. It rests unconformably on steeply dipping Carboniferous sandstones and siltstones. Topographic irregularities in this basement result in rapid thickness variations in the overlying dolomite. The attenuated sequence is also seen in a disused quarry in Limekiln Plantation [3215 5770], Bilton Dene, where 5.3 m of shelly oolitic dolomite with vughs are present.

The attenuated sequences described above thicken rapidly to the south-east, reaching their local maximum east of the River Nidd, beneath Knaresborough. Below Knaresborough Castle, in an inaccessible cliff, approximately 27 m of strata are exposed. The basal 2 to 3 m comprise coarse-grained and reddened Carboniferous sandstone. Above the unconformity there is a diffuse zone of reddish buff sandy dolomite, approximately 6 to 7 m thick, overlain by about 17 m of dolomitic limestone in large-scale cross-bedded units. This sequence and those described below, along Knaresborough gorge, are typical of the Upper Subdivision of the Lower Magnesian Limestone, suggesting that the Lower Subdivision has been largely overlapped in this area. At Bunkers Hill Quarry [3517 5657], large-scale cross-bedding is also well displayed. The quarry shows about 18 m of buff and yellow-buff, medium and thick-bedded dolomite. The beds are cross-stratified, with convex-up cross-stratification in sets, up to 18 m high; these sets commonly make sharp angular contacts against other sets, dipping at an oblique angle, or even in the opposite direction. The palaeocurrent directions were measured by Kaldi (1980a) and are summarised on Figure 14; they show a mainly north-easterly derivation with the sediment deposited on the west-facing side of the Knaresborough oolite bank. The dolomite at Bunkers Hill is mainly porous and recrystallised, but sporadic poorly preserved ooliths are present.

At Abbey Crags [3550 5593], the unconformity between the Lower Magnesian Limestone and the underlying Carboniferous sandstone is very uneven, with buried hills having a relief of about 12 m. The crest of a sandstone hill is exposed on the east side of the gorge; it has a rounded form and is about 32 m wide, with an exposed relief of 8 m (Plate 2). The Carboniferous sandstone is coarse to granule grained, with abundant quartz pebbles; it forms medium and thick cross-bedded units. The sandstone is reddened, but in places it is leached to a buff colour and recemented with dolomite. The contact with the overlying limestone is sharp, but in places small pipes and pockets of weathered material project downwards into the sandstone. The overlying Lower Magnesian Limestone covers the hill with subconcentric drapes of beds 0.1 to 0.15 m thick. These beds are commonly lenticular and show cross-bedding. On the flanks of the buried hill they pass laterally into large-scale cross-bedding on a scale of 1 to 10 m, with sequences wedging in and out. The rock consists of buff-coloured, fine-grained, crystalline to granular dolomite with sporadic poorly preserved ooliths; quartz sand grains and a few pebbles are also present in the basal beds.

South-east from Abbey Crags, the unconformity dips down into a buried valley, but it reappears in a quarry [3570 5574] 100 m east of the Abbey. Here 3.6 m of Carboniferous sandstone are present beneath 13.46 m of Lower Magnesian Limestone. The sandstone is very coarse to granule grained, feldspathic, with quartz pebbles, and occurs in very thick cross-bedded units; the rock is reddened and weathered. At the base of the overlying limestone there is a thin impersistent bed (0 to 0.1 m) of yellow, very coarse-grained sandstone with a dolomitic limestone cement; it is not apparent whether this is part of the Carboniferous sequence, recemented with dolomitic limestone, or a very sandy bed at the base of the Permian. This basal unit is followed by 0.56 m of medium- and thick-bedded, fine-grained crystalline dolomite, with scattered angular quartz grains. The succeeding unit, 1.4 m thick, is of fine- to medium-grained dolomite in thickly laminated to very thin beds, with poorly exposed cross-bedding. This is overlain by 5.45 m of mainly fine-grained crystalline dolomite in thin, medium and thick beds. The upper part of the quarry shows 6.05 m of fine- and medium-grained dolomite; this forms thick and very thick beds passing upwards into medium and thin beds. These upper beds have many water-worn cavities, joints and small caves, some of which contain brecciated, tufa-cemented, collapsed dolomite.

In another quarry, 70 m north of the Priory [3573 5572], the unconformity rises over another buried hill, and 10.5 m of Carboniferous sandstone is exposed beneath the limestone. The sandstone is up to granule grained, with quartz pebbles, and occurs in thick and very thick beds with massive cross-bedding. The unconformity forms a slight bench and the overlying dolomite (c.0.8 m thick) is inaccessible. This is followed by 1.18 m of fine-grained dolomite in thin and medium beds, overlain by 2.9 m of massive, fine- to medium-grained dolomite with calcite-lined vughs. In this upper unit, impersistent traces of bedding are present and numerous water-worn fissures occur.

The unconformity is again exposed at Grimbald Crag [3609 5579], where the footpath runs along it. Here, 29 m of Carboniferous sandstone are present; it is very coarse to granule grained, with abundant quartz pebbles. The sandstone forms very thick beds with large-scale cross-bedding and is considerably reddened. Above the unconformity, the lowest 2 m of beds comprise medium and thick beds of fine-grained dolomite with abundant quartz grains, especially in the lower part. The succeeding unit, 15.5 m thick, is of fine- and medium-grained crystalline dolomite, with vughs; these beds are mainly medium and thick, with large-scale cross-bedding.

The unconformity slopes northwards from Grimbald Crag and crosses the River Nidd south of St Robert's Cave [3611 5609], where it is probably a few metres below the bottom of the exposure. The cave is man-made, with a rectangular front entrance and smooth sides; it extends into the rock face for several metres and once had a lean-to dwelling built in front of it. The face into which the cave is excavated shows 5 m of buff-coloured, fine-grained dolomite with small vughs. Towards the west end of the exposure, the rock consists of a coarse-grained oolite/pisolite with shell fragments. The bedding is mainly horizontal, but cross-bedding is present at the bottom and top of the section. The beds are mainly thin and medium bedded, and a channel structure about 8 m wide is exposed near the bottom of the sequence.

From the vicinity of Abbey and Grimbald Crags, where the Lower Magnesian Limestone is thick, the Limestone thins dramatically eastwards to Goldsborough Mill [3676 5590]. At the mill, the sequence is cut by a fault, upstream of which very coarse-grained Carboniferous sandstone is exposed in the river bed. Overlying this sandstone, about 6 m of dolomite, in thin and medium subhorizontal beds, are present in the west bank. South of the fault, about 8 m of similar dolomite are exposed; here the fault downthrows to the south and the Carboniferous sandstone is not seen. The close proximity of the overlying Middle Marl in exposures upstream [3675 5603], suggests that the Lower Magnesian Limestone is probably only about 10 m thick around here.

*Plompton to North Deighton*

At Guy's Crag [3714 5543], near Goldsborough, the Lower Magnesian Limestone is exposed adjacent to a fault on the west bank of the River Nidd. The sequence appears to be near the top of the formation. It exposes 7.07 m of mainly algal mat-laminated dolomite with granule-sized dolomite fragments, and a few shell fragments in the upper beds. The Lower Magnesian Limestone here appears to be thin; it is overlain nearby to the east by the Middle Marl, and the base is exposed about 0.5 km to the south-west [3697 5484]. Here and near Scalibar Farm [3687 5448], the basal unconformity undulates considerably over buried hills of Carboniferous sandstone.

The basal unconformity is seen again at St Helen's Quarry [3761 5174]. Here, 6 m of reddened and coarse-grained, pebbly, Upper Plompton Grit are overlain by about 2 m of Lower Magnesian Limestone. The contact between the sandstone and limestone is not well exposed, but appears to be uneven. The limestone is dolomitic, oolitic and pisolitic, with rolled shell fragments; it occurs in uneven and lenticular thin and medium beds, in places splitting into very thin and laminated beds.

Newsome Bridge Quarry [3790 5144] (Plate 4) exposes a buried hill of Carboniferous sandstone capped by an inferred Lower Magnesian Limestone patch reef and its flanking deposits (Smith, 1974b and in press). The Upper Plompton Grit here is only slightly reddened and comprises massive, thick-bedded, coarse- to granule-grained, feldspathic sandstone with quartz pebbles; up to 4.5 m of beds are exposed. The contact between the limestone and the sandstone follows the surface of a buried hill with an exposed relief of about 2.5 m. Patches of breccia, up to 1 m thick, occur on both flanks of the hill. The breccia is of local sandstone, in tabular blocks up to 0.9 m long, cemented together with dolomite. The Lower Magnesian Limestone reaches a maximum exposed thickness here of about 7.5 m. The inferred patch reef forms an irregularly shaped, poorly bedded mass which caps the sandstone hill. The reef dolomite is recrystallised and no original reef-building organisms have been identified in situ, but abundant bryozoan fragments, probably reworked from the reef, occur in the surrounding beds. The contact between the inferred reef and the flanking and overlying beds is sharp. The reef is surrounded by thin-bedded, oolitic and pisolitic dolomite with dolomite fragments and shell debris, locally abundant, mainly of *Bakevellia binneyi* and *Schizodus obscurus*. Newsome Bridge Quarry typifies the Lower Subdivision of the Lower Magnesian Limestone as seen farther south (Smith, 1974a and b).

### Middle Marl (Edlington Formation; EZ1–2)

The Middle Marl (Edlington Formation) represents part of the EZ1 and all of the EZ2 cycles of Zechstein deposition (Figure 11). However, near outcrop, the evaporites in this sequence have commonly dissolved, so that only the insoluble red-brown calcareous mudstones crop out. The formation occurs in low ground between the Lower and Upper Magnesian Limestones, as well as around Knaresborough and Harrogate, where it infills the relic topography over and around the Lower Magnesian Limestone oolite bank. Where not drift covered, the Middle Marl weathers to a heavy, red-brown clay soil, which, with its subdued topographic expression, is com-

monly the only evidence for its presence. However, the formation crops out along the Nidd valley at Knaresborough and Little Ribston where, mainly in landslipped ground, red-brown silty mudstone is exposed; near Goldsborough [3784 5467], a few fragments of gypsum also occur in these beds.

The Middle Marl, near and at outcrop, is usually 10 to 15 m thick, but in places it thins to 5 m due to cambering of the overlying Upper Magnesian Limestone. It also thins to a feather edge over the Harrogate Anticline [337 555], where it overlaps the Lower Magnesian Limestone to rest directly on Carboniferous strata. Boreholes show that the Middle Marl generally ranges in thickness from

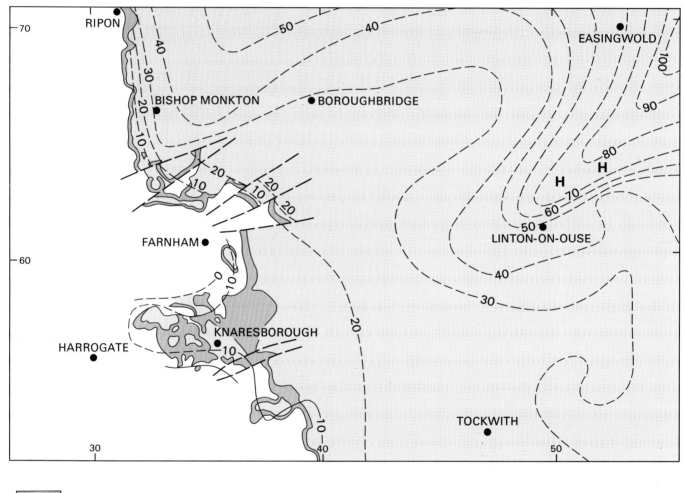

Outcrop of Middle Marl

Middle Marl concealed by younger deposits

Scale 0  1  2  3  4  5km

—10— Isopachytes of Middle Marl, contour interval 10m

**Figure 15**  Isopachyte map for the Middle Marl and its down-dip equivalents in the Harrogate district. H denotes the presence of halite beds in the Middle Marl; these halite beds coincide with the place where the Middle Marl is thickest and they may have been removed by dissolution from some of the remaining areas in the east of the district.

10 to 40 m, with a maximum of 100 m in the north-east of the district (Figure 15). Most of this variation is due to the presence, or preservation, of the ?Hayton Anhydrite at its base, but the thick sequence in the north-east may be explained by the presence of halite.

The ?Hayton Anhydrite is up to 18 m thick, and is composed mainly of bluish grey anhydrite with a slightly nodular texture, picked out by buff-coloured dolomite; it represents the evaporite phase of the lowest Zechstein cycle (EZ1). Although probably originally deposited as gypsum in a hypersaline lagoon environment Smith (1989), the evaporite beds were altered to anhydrite during subsequent burial and diagenesis. However, anhydrite is metastable, hydrating to gypsum if subsequent exhumation brings it into contact with groundwater at depths less than 1000 m (Murray, 1964, Mossop and Shearman, 1973). The Hayton Anhydrite consequently passes westwards, updip, into an irregular belt of secondary gypsum running subparallel to the Middle Marl outcrop. On continued contact with passing groundwater the gypsum progressively dissolves, causing the overlying strata to founder (Smith, 1972; James et al., 1981; Cooper, 1986). This potential hazard is discussed further in chapter eight.

The ?Hayton Anhydrite is overlain by grey and red calcareous and gypsiferous mudstones up to 8 m thick. These are in turn overlain by a thin lenticular dolomitic limestone, possibly the Kirkham Abbey Formation (Smith et al., 1974), which forms the carbonate phase of the second Zechstein cycle (EZ2). This unit does not crop out, but is known from boreholes, mainly in the east of the district, to be 0 to 5 m thick. It is also present in the fault-controlled trough proved by the Spellow Hill Borehole (Appendix 1), where 5.34 m of 'limestone' is recorded within the Middle Marl. The upper 15 to 20 m of the Middle Marl are composed dominantly of laminated to thinly bedded, mainly red-brown, siltstone and mudstone, with sporadic thin gypsum and anhydrite beds. These beds are the terrigenous and marginal-marine lateral equivalents of the Fordon Evaporites (Stewart, 1963; Taylor and Colter, 1975; Colter and Reed, 1980; Smith, 1980), which represent the halite phase of the EZ2 cycle, and which are confined to the more central part of the Zechstein basin, farther east (Figure 11).

**Details**

Southwards from Ripon, the Middle Marl is not exposed for about 7 km, to Burton Leonard. A water borehole near Burton Leonard (Appendix 1) proved the Middle Marl to be 27.73 m thick, mainly red marl with limestone and gypsum beds; some of the 'limestone' beds may be massive gypsum, and one of them is taken to equate with the Kirkham Abbey Formation. About 1 km south of the borehole the Middle Marl forms a steep slope [331 635] between outcrops of the Lower and Upper Magnesian Limestones; here the sequence is about 6 to 8 m thick. The dramatic thinning of the sequence may be explained by dissolution of the gypsum beds and cambering of the Upper Magnesian Limestone.

The full sequence of Middle Marl is also proved in the Spellow Hill Borehole (Appendix 1). Here, the marl sequence is 22.03 m thick, with a 5.34 m thick limestone (possibly equivalent to the Kirkham Abbey Formation) near the middle of the sequence. There is also an 18.36 m thick unit of 'limestone marl' recorded below the marl; this is probably gypsum, but could alternatively be part of the Lower Magnesian Limestone.

From Spellow Hill to Knaresborough gorge the Middle Marl is not exposed. At Scotton Banks [3317 5807], approximately 15 m of poorly exposed and landslipped brownish red, purple, yellow and grey mudstone are present above the Lower Magnesian Limestone. The Middle Marl also crops out on either side of the Nidd gorge, where the Lower Magnesian Limestone is attenuated, south-eastwards almost to the railway viaduct, but the exposures are mainly weathered to a red-brown soil. A small outlier of the Middle Marl was also proved in a trench along Knaresborough High Street [3512 5702]. The marl is also seen in the railway cutting near Starbeck [3404 5629; 3380 5595] and is present in numerous small exposures south of the River Nidd, eastwards to near Goldsborough Mill [3700 5584] where it forms landslipped ground (Plate 12).

South of Goldsborough Mill, the Middle Marl forms a steep, partly landslipped bank beneath the escarpment of the Upper Magnesian Limestone. Here [3745 5553] 4 to 5 m of laminated and thinly bedded, hard, silty, red mudstone with green reduction spots are exposed. Around Little Ribston [3846 5383], the marl forms a belt of landslipped ground below the Upper Magnesian Limestone and an area of heavy red clay soil around Throstle Nest [3737 5335]. Southwards from Little Ribston, the Middle Marl is mapped on the basis of its topographical expression and is proved in a few exploration boreholes.

**Upper Magnesian Limestone (Brotherton Formation; EZ3)**

The Upper Magnesian Limestone (Brotherton Formation) represents the carbonate phase of the third English Zechstein cycle (EZ3). It forms a low, moderately marked escarpment east of the Lower Magnesian Limestone and Middle Marl. Around Knaresborough and Harrogate, however, it overlaps these formations westwards and locally rests directly on Carboniferous strata [339 548]. The Upper Magnesian Limestone is exposed in a few quarries and outcrops, but much of it has been mapped from the distribution of lithologically and faunally distinctive field brash. It is characteristically composed of thinly bedded, whitish or pinkish grey, porcellanous, calcitic dolomite and dolomite that produce a fetid smell when broken; it is commonly cross-bedded. Typical exposures are near Burton Leonard [3319 6345], where 6 m of beds are seen (Plate 7), and in the Knaresborough gorge [3567 5552] (Burgess and Cooper, 1980a). A small quarry in the grounds of Ribston Hall [3900 5400] also exposes the formation, where it is completely recrystallised and composed of a coarsely crystalline dolomite.

The Upper Magnesian Limestone represents a renewed marine incursion which flooded the flat coastal sabkha plain of the Middle Marl. Consequently, the formation is laterally persistent and relatively uniform in thickness, ranging from 8 to 12 m at outcrop. Down dip the formation gradually thickens; in the east of the district boreholes have proved between 21 and 28 m of beds (Figure 16). The restricted and distinctive biota, which typifies the formation, includes *Calcinema permiana*, an alga, with the bivalves *Liebea squamosa* and *Schizodus*

**Plate 7**    Upper Magnesian Limestone near Burton Leonard [3319 6345]. Typical thin and very thinly bedded, white-coloured, dolomitic limestone (L 1801).

*obscurus.* The alga takes the form of thread-like tubes with a diameter of between 0.5 and 1.0 mm; it is so abundant that it sporadically forms small lag deposits and rippled marked drifts of material. *Calcinema permiana* in abundance is almost restricted to, and diagnostic of, the carbonate phase of the third (EZ3) Zechstein cycle (Pattison et al., 1973).

At and near the outcrop, foundering of the Upper Magnesian Limestone, caused by the dissolution of gypsum derived from the thick ?Hayton Anhydrite in the Middle Marl, is a common phenomenon (Smith, 1972; Cooper, 1986, 1989). Exposures of the limestone therefore often have steeper, and directionally more variable dips than those in the underlying Lower Magnesian Limestone; local cambering on the mudstones of the Middle Marl also affects the dip. Dips of 15° occur in the Burton Leonard quarry [3319 6345] because of foundering; cambering and foundering elsewhere locally produce dips of up to 30° in all directions and are the cause of the very localised angular open folds, with a wavelength of a few metres, in the Ribston Hall quarry [3900 5400]. Foundering of the Upper Magnesian Limestone

north and west of Knaresborough, combined with the abrupt westward fall of the base of the Middle Marl, brings the top surface of the limestone down to a similar topographical level as that of the top of the Lower Magnesian Limestone, even though the regional dip is to the east.

**Details**

From Ripon southwards to near Burton Leonard, the Upper Magnesian Limestone is not exposed, except for small areas of field brash [3240 6510]. South-east of Burton Leonard, it is exposed in a disused quarry [3319 6345], where 6 m of beds are present. The limestone is dolomitic, grey and white, with sporadic pink staining. It forms uneven and lenticular, mainly very thin and thin beds with a few medium beds; partings of red clay are common. The rock is generally porcellanous, but it is also partly dolomitised and contains numerous small vughs. Abundant remains of the alga *Calcinema permiana* are commom, drifted together in small channel deposits and as ripple-drift lags. In addition to *Calcinema*, the remains of *Liebea squamosa* and *Schizodus obscurus* are also common. The beds at this quarry have directionally variable dips caused by the dissolution of the underlying Middle Marl gypsum.

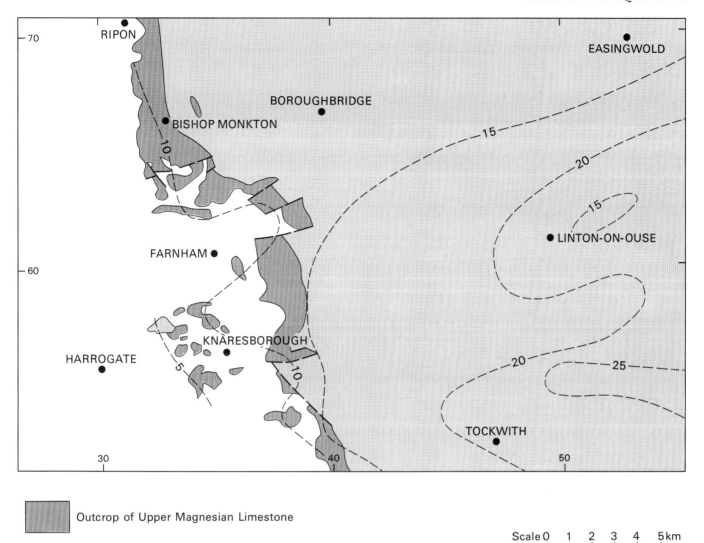

Outcrop of Upper Magnesian Limestone

Upper Magnesian Limestone concealed by younger deposits

Scale 0   1   2   3   4   5 km

—10— Isopachyte of Upper Magnesian Limestone, contour interval 5m

**Figure 16**    Isopachyte map for the Upper Magnesian Limestone in the Harrogate district.

Near Copgrove, 4 m of the Upper Magnesian Limestone are exposed in a disused quarry adjacent to a glacial drainage channel [3463 6298]. The dolomitic limestone here is porcellanous and thinly bedded, with ripple-drift cross-lamination; at about the middle of the section, a 0.3 m bed of dolomitic mudstone is also present. The beds are abundantly fossiliferous with the same species present as at Burton Leonard (see above). Southwards from here, the limestone is seen in several places as field brash, but is not properly exposed again until near Scriven where, in a small garden exposure, red mudstone (Upper Marl) is present overlying dolomite.

In Knaresborough gorge, the Upper Magnesian Limestone is exposed at Bilton Park, 300 m west of Bilton Hall [3311 5746]. Here, approximately 5 m of pale grey and white dolomitic limestone with vughs occurs in laminated, thin and very thin beds. *Calcinema*, *Liebea* and *Schizodus* are all present. On the west side of Knaresborough gorge, near Birkham Wood [3567 5552], 1.5 m of the Upper Magnesian Limestone are exposed. The rock here is white and pale grey porcellanous dolomite with vughs. It occurs in thin and uneven beds and has yielded *Calcinema permiana*, *Liebea* sp. and *Schizodus obscurus*. The limestone is also poorly exposed in the railway cutting near Harrogate Golf Course [3427 5650], where it is seen to overlie red mudstone of the Middle Marl.

At Goldsborough, the Upper Magnesian Limestone is exposed in a small disused quarry [3805 5624], where 0.5 m of thin-bedded white limestone is exposed. The formation also crops out along the edge of the escarpment overlooking the River Nidd; here, 1.8 m of soft thin-bedded dolomite with the remains of *Calcinema* are present.

In the ornamental grounds of Ribston Hall [3900 5400], an extensive disused quarry exposes 2.6 m of the Upper Magnesian Limestone. The limestone here is dolomitic, buff and white, and pink in places; it is coarsely crystalline, with abun-

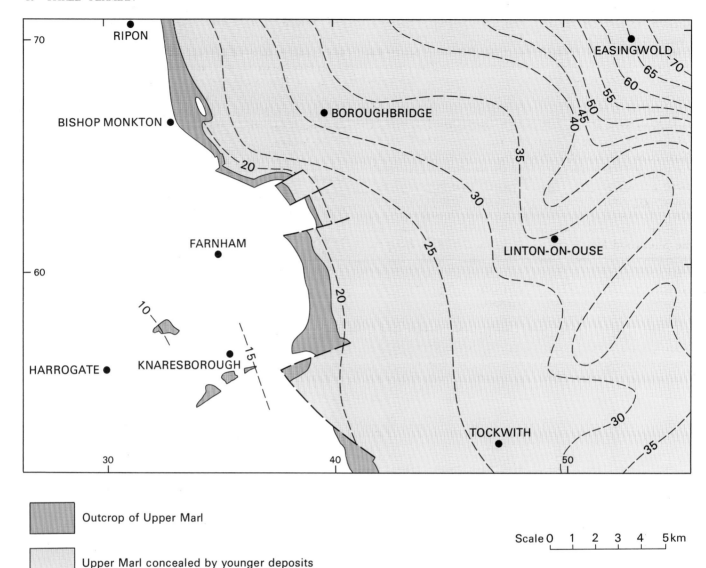

**Figure 17**  Isopachyte map for the Upper Marl and its down-dip equivalents in the Harrogate district.

dant vughs. The bedding is mainly very thin and thin with sporadic medium beds, and a few traces of cross-lamination are present. The dip of the beds is variable, up to 20°, with the beds forming sharp angular folds of 3 to 4 m wavelength. The folds have been produced by foundering, after the dissolution of gypsum in the Middle Marl.

On the west side of the River Nidd, the Upper Magnesian Limestone forms another escarpment capping the hill west of Little Ribston. On the eastern edge of this, overlooking the river, the limestone is exposed in small quarries. In one of these [3863 5377], 3.5 m of pale grey, white and buff dolomitic limestone are exposed. The rock is porcellanous to very fine grained and powdery, with sporadic vughs; it occurs in thickly laminated, thin and medium beds, some of which wedge out. The dip in the exposure is variable, up to 10° in most directions; this is probably due to foundering.

## Upper Marl (Roxby Formation; EZ3–5)

The Upper Marl (Roxby Formation) is not generally exposed in the district, but is present in a tract of low, drift-covered ground lying between the escarpments of the Upper Magnesian Limestone and the Sherwood Sandstone Group. However, south and west of Knaresborough, the Upper Marl does crop out along the Nidd gorge in the cores of a graben and a small syncline. Where the marl is (exceptionally) exposed, it consists of brownish red, silty mudstone. It weathers to a heavy red-brown soil and is often landslipped.

The Upper Marl thickens eastwards into an evaporite sequence, with anhydrite, halite and sylvinite. At outcrop the unit is between 10 and 18 m thick, but down dip it thickens to between 30 and 40 m in the east of the district, and reaches more than 70 m thick in the north-east (Figure 17). Much of this increase in thickness is due to the presence of interbedded third cycle (EZ3) evaporites, mainly the Billingham Main Anhydrite at the base overlain by the Middle Halite and Potash. Thin fourth and fifth cycle evaporites, including the Upper and Top Anhydrite beds along with the Upper Halite and Potash (Figure 11), are also present in the Upper Marl, which thus spans an interval from below the top of cycle EZ3 to the top of EZ5. Like the anhydrite in the Middle Marl, these anhydrite beds pass westwards into zones of secondary gypsum, which in turn pass into dissolution zones where the overlying beds are commonly foundered (see chapter eight); this subsidence affects strata well up into the Sherwood Sandstone Group (Smith, 1972; Cooper, 1986, 1989). The anhydrite/gypsum beds in the Upper Marl are commonly banded, with traces of dolomite, and probably formed as lagoonal and sabkha deposits on the Upper Magnesian Limestone platform.

The junction between the Upper Marl (Saliferous Marl in the east; Figure 11) and the overlying Sherwood Sandstone Group is an interbedded gradation over a few metres as the lithology becomes more arenaceous. The contact is almost certainly diachronous and at outcrop, near the western margin of the basin, it marks a transition to completely terrigenous sedimentation. This part of the sequence is unfossiliferous, but near the outcrop the boundary between the Permian and Triassic is thought to lie in the lower part of the Sherwood Sandstone Group (Smith et al., 1974). However, to the east of the district, in the main North Sea basin, the boundary is probably within the Saliferous Marl (Pattison et al., 1973, p.223; Warrington et al., 1980).

### Details

The only extensive exposure of the Upper Marl is at Bilton Park, 300 m west of Bilton Hall [3308 5746]. Here, the underlying Upper Magnesian Limestone and the overlying Sherwood Sandstone are moderately exposed; between them, approximately 10 m of red silty mudstone are present. In the faulted graben south of Knaresborough the Upper Marl also crops out, but here it is only seen as a heavy red clay soil in the fields. Along most of the outcrop, east of the Upper Magnesian Limestone escarpment, information on the Upper Marl is derived mainly from confidential exploration boreholes.

No fossils are known from the formation in the district; material from the unit at 40.47 m depth in the Claro House Borehole [4074 5999] was examined for palynomorphs, but proved barren.

FOUR

# Triassic

With the onset of the Triassic period, the western part of the Zechstein Basin established in the Permian became the site of continental redbed sedimentation. A thick sequence of mainly red sandstones was succeeded by marls (red, variably dolomitic mudstones and siltstones) with evaporites. Towards the end of the period, marine conditions spread across the area and led to the deposition of black shales, greenish calcareous mudstones and dark grey calcareous mudstones with thin limestones.

The Triassic redbed sediments have traditionally been divided into the Bunter and the overlying Keuper, names adopted from the Triassic of Germany (Sedgwick, 1829). On the original (1884) one-inch geological map of the district, the lower part of the red sandstone was classified as Bunter Sandstone and the upper part as Keuper Sandstone (Figure 18). However, boundaries were drawn neither between these two formations, nor to delimit the lower and upper boundaries of the overlying Keuper Marl mudstones. The succeeding marine strata were referred to as the Rhaetic Beds by comparison with sequences elsewhere, and although symbolised on the one-inch map as part of the Jurassic, were included in the Triassic by Fox-Strangways (1908). However, on the 1959 reprint of the Harrogate sheet, the boundary between the Keuper Sandstone and Keuper Marl is delineated. Later workers recognised that in Yorkshire no subdivision could be made within the arenaceous unit, which was then assigned solely to the Bunter Sandstone (Figure 18; Warrington, 1974).

In 1980, Warrington et al. concluded that the traditional Triassic subdivisions not only differed significantly in terms of age and facies from the European units with which they were supposed to correspond, but were also inconsistent with modern stratigraphical terminology.

| PERIOD | STAGE | | LITHOSTRATIGRAPHY | | | | |
|---|---|---|---|---|---|---|---|
| | | | PRIMARY SURVEY 1874 (Fox-Strangways, 1908) | | (Warrington, 1974; Hemingway, 1974 and Smith, 1974a) | | PRESENT NOMENCLATURE (after Warrington et al. 1980; Powell, 1984 and Smith et al. 1986) |
| JURASSIC | HETTANGIAN | | LOWER LIAS | | LOWER LIAS | CALCAREOUS SHALES | LIAS GROUP | REDCAR MUDSTONE FORMATION |
| TRIASSIC | RHAETIAN | | RHAETIC BEDS | WHITE SHALES | RHAETIC | | PENARTH GROUP | LILSTOCK FORMATION (COTHAM MEMBER) |
| | | | | BLACK SHALES | | | | WESTBURY FORMATION |
| | | | | TEA GREEN MARL | | TEA GREEN MARL | | BLUE ANCHOR FORMATION |
| | NORIAN | ? | KEUPER MARL | | KEUPER MARL | | MERCIA MUDSTONE GROUP | |
| | CARNIAN | ? | | | | | | |
| | LADINIAN | ? | | | | | | |
| | ANISIAN | ? | | | | | | |
| | SCYTHIAN | ? | KEUPER SANDSTONE | | BUNTER SANDSTONE | | SHERWOOD SANDSTONE GROUP | |
| | | ? | BUNTER SANDSTONE | | | | | |
| PERMIAN | | ? | UPPER MARLS | | UPPER MARLS | | ROXBY FORMATION | |

**Figure 18**  Stratigraphical framework for the Triassic strata in the district, comparing the previous nomenclature with that of the present.

The formal lithostratigraphical terms introduced by these authors are adopted here: the arenaceous beds (Bunter and Keuper Sandstone) are assigned to the Sherwood Sandstone Group, the overlying marls (Keuper Marl) to the Mercia Mudstone Group, and the succeeding marine shales and calcareous mudstones (Rhaetic Beds) to the Penarth Group. The current and former classifications of the Triassic strata in the district, and their chronostratigraphical divisions, are shown in Figure 18.

The appearance of ammonites of the genus *Psiloceras* is used to recognise the Triassic–Jurassic boundary in Britain (Cope et al., 1980; Warrington et al.,1980). *Psiloceras* typically appears a few metres above the base of the Lias Group, the lowermost beds of which are thus assigned a Triassic age (Figures 18 and 19); in this account these lowermost beds are described, for convenience, with the remainder of the Lias Group, in chapter five.

## SHERWOOD SANDSTONE GROUP

The group consists predominantly of poorly cemented, unfossiliferous, fine- to medium-grained sandstones, mainly brick-red but also grey, yellow and mottled. In general, they are medium- to thick-bedded and commonly cross-stratified; small-scale ripple marks are evident on some bedding planes. Clasts consist mainly of quartz but also include feldspars and rock fragments; ilmenite and zircon are the dominant heavy minerals (Smithson, 1931). Mudstone-flake breccias are common, but there is no evidence for the presence of the coarse pebbly facies found farther south (Warrington, 1974). Thin beds of red siltstone and mudstone occur throughout the group, but are concentrated mainly near the transitional boundary with the underlying Permian Upper Marl. Typically, they form the tops of upward-fining cycles, 1 to 2 m thick, which commence with an erosively based mudstone-flake breccia, continue with a cross-stratified sandstone, commonly with mudstone flakes, and grade into siltstone and mudstone. These last show sporadic evidence of subaerial exposure, with mudcracks and gypsum epimorphs. The characteristics of these cycles suggest deposition under distal fluvial conditions, and the Sherwood Sandstone of the district may have been laid down on an extensive alluvial plain, the head of which probably lay far to the south.

Strata of the Sherwood Sandstone Group underlie much of the district, but are largely concealed by drift. However, considerable information is available from boreholes which have been drilled to exploit their potential as an aquifer, and from coal exploration boreholes. Such data, at present confidential, indicate that the thickness of the group increases from some 300 m in the north of the district to approximately 400 m in the centre and south. The lower beds are sporadically exposed along north-north-west-trending ridges, which extend from north of Boroughbridge to south-east of Tockwith, and may represent the scarp features of slightly more resistant horizons. The sandstones appear mainly as field brash around Little Ouseburn [449 606], Green

Hammerton [458 568] and Long Marston [490 515], but are also seen in a few disused quarries.

No fossils are known from the group in the district; material from the lowermost beds at 13.00 m depth in the Claro House Borehole [4074 5999] was examined for palynomorphs, but proved barren. There is, therefore, no direct biostratigraphical evidence for the age of the group (Figure 18), which is constrained only by its stratigraphical position between late Permian deposits and the Mercia Mudstone Group, the lower beds of which have yielded Anisian (Middle Triassic) palynomorphs in the adjacent Thirsk district (Powell et al., in press).

### Details

At Aldborough, an old quarry [4030 6619] exposes 8 m of red-brown and yellow, trough cross-stratified, medium-grained sandstone. Another quarry nearby [4043 6607] shows 2.3 m of similar beds. In the railway cutting [4543 5587] some 750 m east of Cattal Station, 7 m of similar red-brown fine to medium grained, cross-bedded sandstones are patchily exposed; a lens of purple-brown mudstone is also present. The cross-bedded, units up to 1.3 m thick, show evidence of channelling and erosion; red-brown mudstone fragments occur at the base of one unit Near Tockwith, 1.45 m of the Sherwood Sandstone is exposed in the south bank of the River Nidd [4420 5283]. Here the sandstone is orange-red and coarse grained, and occurs in thick cross-stratified beds.

## MERCIA MUDSTONE GROUP

Around the margins of the depositional basin, the fluvial sedimentation of the Sherwood Sandstone Group was abruptly terminated by minor earth movements, followed by erosion. The strata of the succeeding Mercia Mudstone Group are predominantly argillaceous and evaporitic. Within the district, the outcrop of the Mercia Mudstone is completely masked by drift and there is no field evidence for the initial period of erosion, which has been identified both to the north (Smith, 1970a; Taylor et al., 1971) and to the south (Edwards, 1951; Smith and Warrington, 1971). Borehole records do, however, testify to the abruptness of the change from sandstone to mudstone.

The group, 190 m in thickness, is present in the north-eastern part of the district only. It consists largely of dolomitic mudstone and silty mudstone, commonly brown, red-brown, or red, but including thin greenish brown bands and patches. The mudstones are generally blocky, but may be finely interlaminated with clean-sorted siltstones. Thin beds of grey-green, hard siltstone and very fine-grained sandstone are present sporadically throughout the succession. Gypsum and, at depth, anhydrite occur not only as beds, which rarely attain a thickness of 1 to 2 m, but also as nodules and ramifying veins. Geophysical logs of boreholes show that these evaporitic beds are concentrated at three levels, near the base, middle and top of the group, as they are farther south (Smith, et al., 1973) and east (Whittaker et al., 1985). Brecciation of the mudstones, probably due to collapse after gypsum dissolution, has been noted from boreholes in the Tholthorpe area (Stanczyszyn, 1982).

Regional studies (Warrington, 1974) suggest that the uppermost unit of the group, the Blue Anchor Formation (formerly the Tea Green Marl, Figure 18), comprising 10 m of grey-green dolomitic mudstones, is probably present in the district. Fragments of lithologically similar mudstones occur at this level 1.5 km east of Easingwold [at 5436 6980].

The Mercia Mudstone has yielded no fauna in the district. Its colour and argillaceous nature, the absence of fauna and the presence of evaporites suggest deposition under conditions varying from highly saline lagoonal to terrigenous emergent, with infrequent flash floods bringing coarser-grained material into the basin. Some workers have drawn parallels with playa lakes (Anderton et al., 1979); others, pointing to the presence of a marine carbonate and halite facies in a possible correlative (the Muschelkalk) farther east in the depositional basin (Warrington, 1974), advocate a hypersaline epeiric sea, fringed by sabkhas.

Palynology samples from a borehole at Burn Hall [5386 6566] and an excavation near Easingwold [5436 6987] proved barren. However, material from 4.4 to 4.5 m depth in Easingwold No. 5 Borehole [5551 6731], Greycarr Lodge, yielded Carboniferous spores, assessed as reworked, and pollen, including *Lunatisporites* sp. and *Ovalipollis pseudoalatus*. The presence of *O. pseudoalatus* is indicative of a mid to late Triassic (Ladinian to Rhaetian) age for the beds at this site. In the adjacent Thirsk district, to the north, palynomorph assemblages indicative of Middle Triassic (Anisian and Ladinian) and later Triassic (Carnian or Norian) ages have been recovered from the group (Powell et al., in press).

### Details

There are no exposures of the Mercia Mudstone Group; the only details come from boreholes. North-east of Tholthorpe the group is proved in a water borehole at Spring Head [4864 6960]; this penetrated drift to 15.5 m, shale with gypsum to 20.1 m, red shale to 32.5 m, sandstone (Sherwood Sandstone Group) to 46.3 m and blue shale to the bottom at 49.3 m. A mineral assessment borehole nearby [4879 6928] (Stanczyszyn, 1982) penetrated 0.8 m of dark greyish green, micaceous siltstone with probable gypsum fragments at a depth of 18.6 m, beneath glacial deposits. About 800 m to the north-east, another borehole of the same series [4933 6996] penetrated 1.4 m of reddish brown mudstone with green mottling and beds of siltstone at a depth of 13.4 m. Near Derrings Farm [4761 6858], another water borehole proved the following sequence: drift to 14.6 m, marl with gypsum to 22.9 m and red sandstone (Sherwood Sandstone Group) to the bottom of the hole at 76.2 m. At Station Farm [5053 6739], the lower part of the Mercia Mudstone Group was proved in a water borehole which penetrated drift to 14.7 m, red and green marl with gypsum to 25.0 m and soft red sandstone (Sherwood Sandstone Group) to the bottom of the hole at 41.5 m. A similar part of the sequence was also proved at Burn Hall [5392 6542], Huby, where a water borehole penetrated 10.8 m of drift, overlying red and green marl and sandy marl with gypsum to the bottom of the hole at 20.3 m.

## PENARTH GROUP

The essentially continental, oxidising environment of the Mercia Mudstone Group was followed, rather abruptly, by marine, muddy, reducing conditions during deposition of the Penarth Group. The transition may be reflected also in the grey-green mudstones of the Blue Anchor Formation, which, elsewhere in eastern England, contain fish scales (Kent, 1980). However, the major break lies at the base of the succeeding Penarth Group (formerly Rhaetic; Figure 18).

From evidence outside the district, the Penarth Group is divided into the Westbury Formation below and the Lilstock Formation above (Warrington, 1974; Warrington et al., 1980). The Westbury Formation comprises brown, dark grey or black, sporadically carbonaceous and pyritic shales, with a few thin beds of fine-grained sandstone, commonly bioturbated. The marine benthonic fauna indicates a major marine transgression; reducing conditions existed below the surface of mudstones deposited during this transgression. The succeeding Lilstock Formation is dominantly calcareous and is represented by pale green, soapy-textured mudstones, which have a restricted fauna; in the Harrogate district all these beds are assigned to the Cotham Member of the formation. They represent a phase of lagoonal sedimentation, after which the marine transgression continued with the deposition of the alternating shales and limestones typical of the Redcar Mudstone Formation of the Lias Group (chapter five).

The Penarth Group crops out only in two small drift-covered areas, to the west and east of Easingwold in the north-east of the district, where it is some 9 m thick. The westerly occurrence is inferred from its position between two boreholes, one penetrating the Redcar Mudstones and the other penetrating the Mercia Mudstones. Boreholes immediately south-east of Easingwold proved only Mercia Mudstone in an area that, on the evidence of fragmentary material from a well sinking, was shown as Rhaetic on the primary survey map. The contention of Tate and Blake (1876, p.33) that it was 'improbable that the rocks were there in situ' is thus upheld. Farther east, however, a shallow BGS borehole at Crayke Lodge [5458 6992] penetrated dark grey and brown mudstones of the Westbury Formation, from which Rhaetian palynomorphs have been obtained (Benfield and Warrington, 1988).

### Details

The Crayke Lodge Borehole [5458 6992] proved, beneath 3.2 m of till, 0.7 m of weathered, grey and brown ferruginous mudstone resting on 0.6 m of grey-black, slightly silty and partly fissile mudstone (Benfield and Warrington, 1988). The dark-coloured, predominantly argillaceous nature of the sediments is consistent with a lithostratigraphical correlation with the Westbury Formation. Three palynomorph assemblages from the beds between 3.9 and 4.5 m in this borehole are dominated by spores and pollen, but contain a few organic-walled microplankton. The spore and pollen associations are dominated by specimens of *Ovalipollis pseudoalatus*, *Ricciisporites tuberculatus*, *Rhaetipollis germanicus*, *Classopollis torosus* and *Vesicaspora fuscus*, but include sporadic *Limbosporites lundbladii* and

*Quadraeculina anellaeformis*, and are indicative of a Rhaetian (Late Triassic) age (Benfield and Warrington, 1988). The subordinate organic-walled microplankton association, which comprises acritarchs and possible remains of the dinoflagellate cyst *Rhaetogonyaulax rhaetica*, is indicative of a marine environment.

Material from 4.1 to 4.5 m depth in Easingwold No. 4 Borehole [5552 6902], Park House Farm, at the boundary with the York district to the east, yielded spores and pollen, including *Ovalipollis pseudoalatus* and *Classopollis torosus*, indicative of a late Triassic (Norian? to Rhaetian) age.

FIVE

# Jurassic

In North Yorkshire, marine conditions prevailed throughout early Jurassic times. The Cleveland Basin (Kent, 1980) was established and includes the Jurassic deposits of the Harrogate district, North York Moors, Howardian Hills and Cleveland Hills. The southern margin of the basin parallels the Howardian Hills Fault Belt (Kent, 1980) and approximately marks the edge of the basement blocks established during the Carboniferous. A basal sequence of relatively deep-water, predominantly argillaceous sediments was followed by shallower water sandstones, ironstones and mudstones which are in turn overlain by deeper-water mudstones. These three lithostratigraphical units were termed the Lower, Middle and Upper Lias respectively, the Middle Lias originally being divided, where possible, into a 'Sandy Series' and an 'Ironstone Series' (Fox-Strangways, 1908; Fox-Strangways et al., 1886; Fox-Strangways and Barrow, 1915). The Lias sequence was subsequently reclassified by Hemingway (1974) and Cope et al. (1980). Within the last decade, it has been redefined (Powell, 1984; Knox, 1984) as the Lias Group and subdivided into five mappable formations. These, with their chronostratigraphical classification and the equivalents in previous nomenclature schemes are shown in Figure 19. The upper part of the Whitby Mudstone Formation and the overlying Blea Wyke Sandstone Formation do not occur within the district.

Elsewhere in the Cleveland Basin, particularly on the coast, the Jurassic has also been subdivided biostratigraphically, principally on the basis of its ammonite faunas (Cope et al., 1980). No ammonites have been found in the poorly exposed rocks of the district. However, studies of the bivalve macrofauna, the calcareous microfauna and the palynomorphs have provided limited information on the biostratigraphy of parts of the sequence. Comparison with the Felixkirk borehole (Powell, 1984; Ivimey-Cook and Powell, 1991), some 17 km to the north-north-west of Easingwold, suggests that in the Harrogate district the lowest few metres of the Lias Group are of Triassic age (Figures 18 and 19).

## LIAS GROUP

Strata of the Lias Group are confined to a small area near Easingwold, in the extreme north-east of the district; four formations are present at outcrop.

### Redcar Mudstone Formation

The formation is poorly known in the district. However, in the Thirsk district to the north, it comprises some 200 m of mainly dark grey mudstones and siltstones; subsidiary thin limestones (concentrated in the lower third),

very fine-grained sandstones, and sideritic and 'chamositic' ironstone beds and nodules are also present (Powell, 1984). In adjacent areas, it generally contains a rich fauna of gastropods, bivalves, ammonites and belemnites and is interpreted as having been deposited in an extensive shelf sea, which varied in depth from relatively shallow to moderately deep.

The outcrop of the formation within this district is entirely drift covered. Evidence is limited to a single, shallow BGS borehole at Primrose Hill Farm, Easingwold [5108 7033], which penetrated, beneath drift, 0.9 m of brown and grey-green calcareous mudstones and siltstones. Palynological preparations from 5.6 to 6.3 m depth in this borehole yielded assemblages dominated by spores and pollen of terrestrial origin, but including tasmanitid algae, *Botryococcus* colonies and acritarchs. The spore and pollen associations include *Cyathidites minor, Osmundacidites wellmanii, Baculatisporites* sp., *Lycopodiacidites rugulatus, Retitriletes* spp., *Perinopollenites elatoides, Tsugaepollenites mesozoicus, Quadraeculina anellaeformis, Alisporites* spp., *Chasmatosporites magnolioides* and *Classopollis torosus*. These associations are of early Jurassic age, but are not older than Sinemurian. The acritarchs are indicative of a marine environment. Also present in the assemblages are taeniate bisaccate pollen of Permian and Triassic age. The presence of these specimens implies reworking through erosion of Permian or Triassic deposits during early Jurassic times.

Tate and Blake (1876) reported that excavations in the 'lower portions' of Easingwold encountered blue clays from which they obtained a fauna of gastropods, bivalves and belemnites which they considered to be of *jamesoni* Zone age (early Pliensbachian). However, recent boreholes in this part of Easingwold indicate that the bedrock is Mercia Mudstone, and so either the fossils were derived from a large erratic within the drift, or, if from the solid, they might indicate a small wedge of Redcar Mudstone caught up in the major fault which passes to the north of Easingwold Town Hall [5288 6983].

### Staithes Sandstone Formation

This formation comprises approximately 20 m of fossiliferous, micaceous, calcareous, very fine- to fine-grained sandstone, yellow-brown at outcrop but grey when fresh, with interbedded grey siltstone and silty mudstone. The sandstones exhibit cross-stratification, ripple cross-lamination and bioturbation. An abundant bivalve fauna, including *Oxytoma inequivalvis, Protocardia truncata* and species of *Pseudopecten*, locally produces conspicuous shell beds in the surrounding districts. The formation represents deposition under shallower-water conditions (with increased wave action and sorting) than those of the preceding Redcar Mudstones.

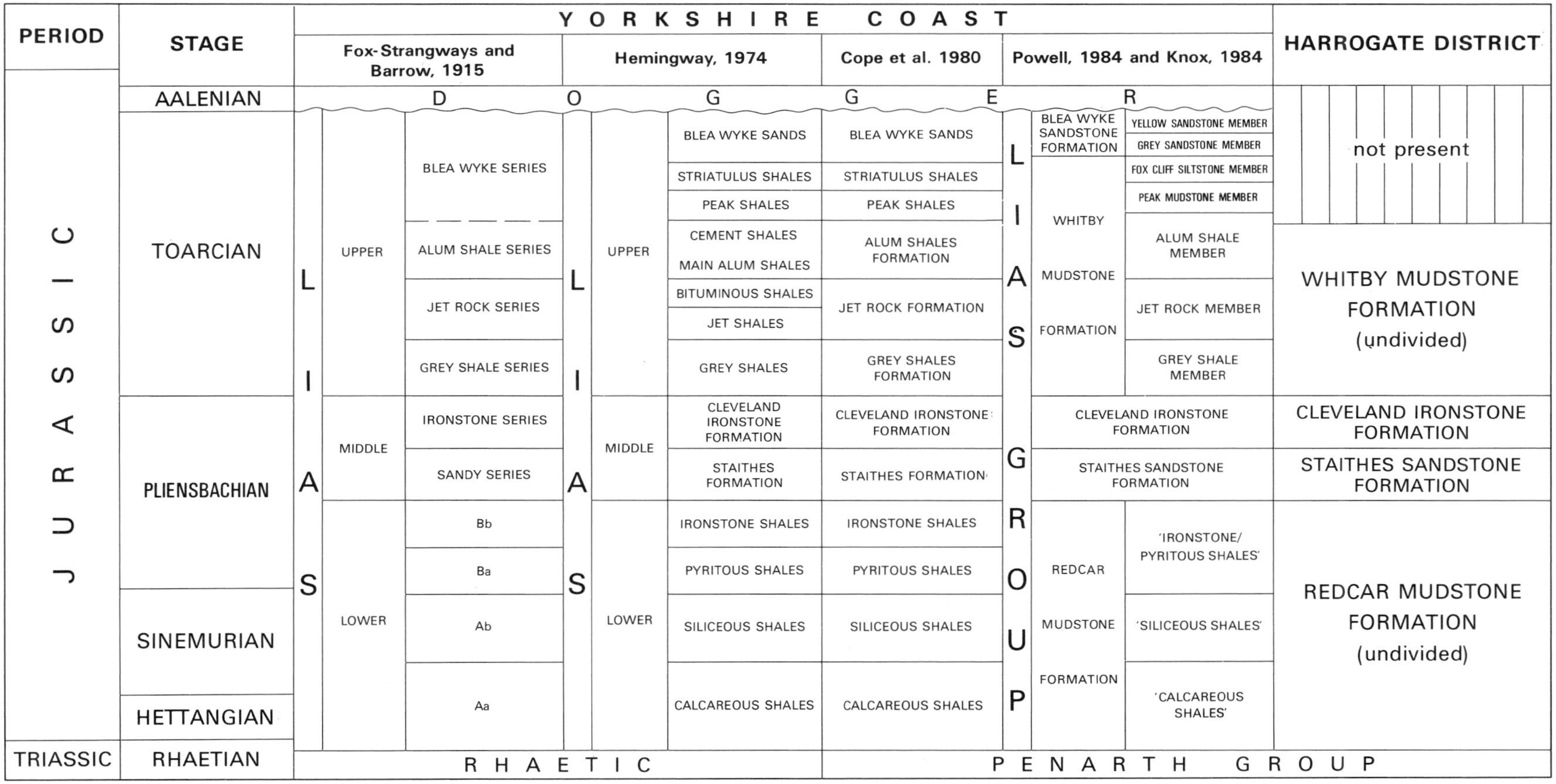

| PERIOD | STAGE | YORKSHIRE COAST — Fox-Strangways and Barrow, 1915 | Hemingway, 1974 | Cope et al. 1980 | Powell, 1984 and Knox, 1984 | HARROGATE DISTRICT |
|---|---|---|---|---|---|---|
| JURASSIC | AALENIAN | BLEA WYKE SERIES (UPPER LIAS) | BLEA WYKE SANDS | BLEA WYKE SANDS | BLEA WYKE SANDSTONE FORMATION — YELLOW SANDSTONE MEMBER / GREY SANDSTONE MEMBER | not present |
| JURASSIC | TOARCIAN | ALUM SHALE SERIES | STRIATULUS SHALES | STRIATULUS SHALES | FOX CLIFF SILTSTONE MEMBER / PEAK MUDSTONE MEMBER (WHITBY MUDSTONE FORMATION) | WHITBY MUDSTONE FORMATION (undivided) |
| JURASSIC | TOARCIAN |  | PEAK SHALES | PEAK SHALES |  |  |
| JURASSIC | TOARCIAN |  | CEMENT SHALES / MAIN ALUM SHALES | ALUM SHALE FORMATION | ALUM SHALE MEMBER |  |
| JURASSIC | TOARCIAN | JET ROCK SERIES | BITUMINOUS SHALES / JET SHALES | JET ROCK FORMATION | JET ROCK MEMBER |  |
| JURASSIC | TOARCIAN | GREY SHALE SERIES | GREY SHALES | GREY SHALES FORMATION | GREY SHALE MEMBER |  |
| JURASSIC | PLIENSBACHIAN | IRONSTONE SERIES (MIDDLE LIAS) | CLEVELAND IRONSTONE FORMATION | CLEVELAND IRONSTONE FORMATION | CLEVELAND IRONSTONE FORMATION | CLEVELAND IRONSTONE FORMATION |
| JURASSIC | PLIENSBACHIAN | SANDY SERIES | STAITHES FORMATION | STAITHES FORMATION | STAITHES SANDSTONE FORMATION | STAITHES SANDSTONE FORMATION |
| JURASSIC | PLIENSBACHIAN | Bb (LOWER LIAS) | IRONSTONE SHALES | IRONSTONE SHALES | 'IRONSTONE/ PYRITOUS SHALES' (REDCAR MUDSTONE FORMATION) | REDCAR MUDSTONE FORMATION (undivided) |
| JURASSIC | SINEMURIAN | Ba | PYRITOUS SHALES | PYRITOUS SHALES |  |  |
| JURASSIC | SINEMURIAN | Ab | SILICEOUS SHALES | SILICEOUS SHALES | 'SILICEOUS SHALES' |  |
| JURASSIC | HETTANGIAN | Aa | CALCAREOUS SHALES | CALCAREOUS SHALES | 'CALCAREOUS SHALES' |  |
| TRIASSIC | RHAETIAN | RHAETIC | | | PENARTH GROUP | PENARTH GROUP |

**Figure 19**  Stratigraphical framework for the Lias Group in the Cleveland Basin, comparing the previous nomenclature with that adopted for the Harrogate district.

The Staithes Sandstone forms a fault-bounded, but largely drift-covered ridge to the north and north-east of Easingwold. In places, the drift is very thin or absent and a brash of sandstone fragments is ploughed up, but no exposures occur.

The formation was proved by the Easingwold No. 12 shallow borehole [5253 7006]. A sparse spore and pollen association from 7.1 to 7.2 m depth in this borehole is indicative only of an early Jurassic age, not older than Sinemurian. Acritarchs from the same depth indicate marine conditions of deposition, but an abundance of plant debris signifies a situation relatively close to land.

The uppermost beds of the Staithes Sandstone at Easingwold may be of the same age as on the Yorkshire coast (cf. Cope et al., 1980, fig. 8B, column PL 17), although that formation was omitted from the correlation column for the Pliensbachian of this general area (Cope et al., 1980, fig. 8B, column PL 15), which shows Ironstone Shales and other units (now Redcar Mudstones) extending up to a level equivalent to the base of the Cleveland Ironstone Formation (Figure 19).

**Cleveland Ironstone Formation**

Within the district, the Cleveland Ironstone comprises some 10 m of greenish grey, silty mudstone and siltstone with at least one bed of dark brown, fossiliferous, oolitic, 'chamositic', sideritic ironstone, 0.6 to 1.0 m thick. Other, thinner ironstone beds may also be present. The depositional environment continued to be shallow marine, with periodic strong wave activity, but iron compounds were precipitated or brought in to a much greater extent.

The formation is almost totally drift covered, but in sporadic small areas abundant angular fragments of oolitic ironstone are ploughed up. One such locality [5422 7029], east-north-east of Easingwold, yielded a prolific fauna including *Tetrarhynchia?* and the bivalves *Oxytoma* (*Palmoxytoma*) *cygnipes, Plicatula spinosa* and *Pseudopecten*, together with indeterminate belemnite fragments. However, none of these fossils provides evidence for the precise biostratigraphical age of the formation. The beds have been penetrated by some wells or boreholes, notably two documented by Tate and Blake (1876, p.142); at Claypenny Hospital [5338 7036] and near Haverwitz Farm [5403 7051], just to the north of the district, they recorded oolitic ironstones about 0.6 m thick. Sparse foraminiferal faunas, comprising long-ranging early Jurassic species, were recovered from the formation from a borehole [5300 7011] in the Easingwold area; the assemblage is dominated by agglutinated species, consistent with unfavourable conditions in a marginal marine situation. A single carapace of the ostracod *Cytherella lindseyensis* was recovered from a sample very close to the top of the formation in the Easingwold No. 10 Borehole [5357 7018], Thornlands, Easingwold. The only published record of this species is from the *margaritatus* Zone (*stokesi* Subzone) of Kirton-in-Lindsey, Lincolnshire (Lord, 1974); consequently, its full range and stratigraphical significance are uncertain. Spores and pollen from 6.87 to 7.30 m depth in the same borehole are indicative only of an early Jurassic age, but are not older than Sinemurian. Sporadic acritarchs from the same depth indicate marine conditions of deposition, but the presence of abundant plant debris signifies proximity to land.

**Whitby Mudstone Formation**

The formation comprises grey to dark grey, sparsely fossiliferous, fissile, locally bituminous mudstones and siltstones. Bands of calcareous concretions are common at some horizons. The formation probably represents a return to quieter, deeper-water marine conditions. The lowest 15 to 20 m of the formation are probably represented within the district, but they are entirely drift covered. At Claypenny Hospital, just to the north of the district boundary, a small roadside cutting [5343 7049] exposes dark grey, finely laminated, fissile mudstones, which yielded a palynomorph assemblage indicative of a Toarcian (possibly early Toarcian) age (Powell et al., in press).

# SIX

# Structure

A principal feature of the folding and faulting in the Upper Palaeozoic and Mesozoic rocks of the district is the reactivation of pre-existing structural lines. It is believed that faults associated with the Acadian earth movements (sensu Soper et al., 1987, formerly Caledonian) remained active during Carboniferous sedimentation, controlling facies distribution and delineating 'blocks' and 'basins' of thinner and thicker deposition as lithospheric extension took place (Lee, 1988; Leeder, 1988; Collinson, 1988; Ebdon et al., 1990). Hercynian movements, which involved a partial inversion of the depositional basins, were accompanied by uplift and erosion, producing the major unconformity that exists between the Carboniferous and Permian rocks. Many of the earlier structures were also reactivated in Jurassic and later times, in response to the stresses associated with the opening of the Atlantic Ocean and development of the Cleveland Basin marginal to the North Sea Basin. On account of this continuity of movement, it is not always possible to determine the age of individual structures or to ascertain how often they have moved.

## STRUCTURES AFFECTING THE CARBONIFEROUS ROCKS

The Carboniferous outcrop in the Harrogate district lies at the eastern edge of a large area of Dinantian and Namurian rocks, where the tectonic pattern is complex, but little studied. A full analysis must therefore await the resurvey of that wider area. The bulk of the structures may be attributed to the climax of the Hercynian orogeny (Asturian phase), but some of them may have been initiated in early Carboniferous times. They form a belt of *en échelon* folding and faulting, which may be traced westwards into the Settle district, and which are attributed to structural inversion resulting from dextral transpression of the Craven Basin sequences against the rigid Askrigg Block (Arthurton, 1983, 1984; Gawthorpe, 1987; Kent 1974).

Within the Harrogate district, the structure of the southern part of the Carboniferous outcrop is dominated by the Harrogate Anticline, the axis of which trends approximately north-eastwards through Harrogate, swinging from a more easterly trend in the south (Figures 3 and 6). The fold is asymmetrical and periclinal, plunging both to the north-east and west. The north-western limb is the steeper, with dips locally up to 80°, and is associated with a high-angle reverse fault which was formerly exposed in the foundation excavations of the Harrogate Conference Centre [3018 5576]. The south-eastern limb is less steep, with dips no greater than 45°. Near the axis, around Beckwith, the Harrogate Roadstone formerly displayed minor folds with axes par-

allel to that of the main fold. Anticlines on similar trends in the Craven Basin to the west were interpreted by Arthurton (1984) as a response to dextral shearing, active from mid-Dinantian to early Namurian times. Exposures are insufficient to determine whether or not these stresses initiated the Harrogate Anticline. However, the involvement of late Namurian and Westphalian rocks shows that comparable stresses were probably active also in late Westphalian times.

To the north, the complementary Farnham Syncline runs east-north-eastwards from Saltergate Hill to Farnham, plunging east (Figure 6). It is poorly exposed, but preserves late Namurian and Westphalian strata at its eastern end before disappearing beneath Permian cover. The northern limb of the Farnham Syncline is cut by the Ripley–Staveley Fault Belt, which also coincides with a marked change in the character of the folding and faulting (Figure 6). The folds to the north of the fault belt are impersistent, of small amplitude and wavelength, and show a variety of axial trends. Among them is the easterly trending and periclinal Cayton Gill Anticline, a near-symmetrical fold with a vertical axial plane, which is cut by the north-westerly trending Scarah Moor Fault. This is an oblique-slip fault which displaces the axial plane of the fold sinistrally, producing a horizontal shift of about 250 m; the downthrow is to the west. Both the folding and wrench faulting are the result of dextral transpression during the Hercynian orogeny; the Scarah Moor Fault may be interpreted as a Reidel shear. The Cayton Gill Anticline, and the associated minor folds present for 2 to 3 km to the north, as far as Haddockstones [2712 6582], probably mark the southern margin of the Askrigg Block. On the block, northwards and westwards from Haddockstones, the dips are low, the structure is simple and the Carboniferous sequence thin.

Adjacent to the Ripley–Staveley Fault Belt, Dinantian rocks were proved beneath the Permian in Ellenthorpe No. 1 Borehole. The borehole was drilled on a gravity high, which has been geophysically modelled by Allsop (1985) as an upstanding area of strata, possibly a horst, related to the fault belt; however, it could possibly be an anticline *en échelon* with the Harrogate Anticline. The Ripley–Staveley Fault Belt has a complex history and it was certainly active in late Carboniferous times. The varying throw along its length and the lack of correspondence of fold structures on either side suggest that it was also partly a line of transcurrent movement, and probably a result of transpression. The fault belt apparently continues to the north-north-east, passing south of Ellenthorpe, where geophysical and borehole evidence suggest that it is represented by a number of pre-Permian faults, with overall downthrow to the south, forming the limit of the North Ouse Coalfield. Farther eastwards, it passes into the fault structures of the Howardian Hills

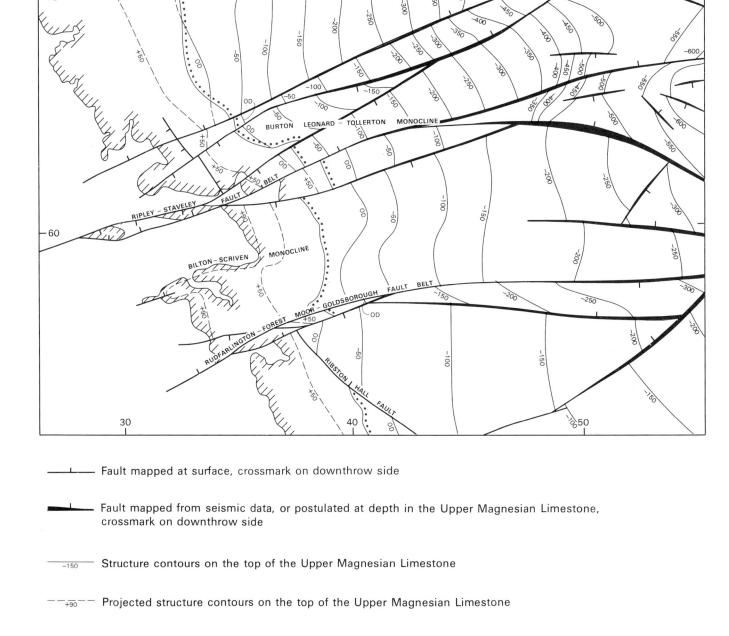

———⊥——— Fault mapped at surface, crossmark on downthrow side

▬▬⊥▬▬ Fault mapped from seismic data, or postulated at depth in the Upper Magnesian Limestone, crossmark on downthrow side

——₋₁₅₀—— Structure contours on the top of the Upper Magnesian Limestone

——₊₉₀—— Projected structure contours on the top of the Upper Magnesian Limestone

•••••••• Top of Upper Magnesian Limestone at outcrop

⁄⁄⁄⁄⁄⁄ Western limit of Permo-Triassic strata

Scale 0  1  2  3  4  5 km

**Figure 20**  Structure contour map for the top of the Upper Magnesian Limestone. West of the top of the Upper Magnesian Limestone outcrop the structure contours are projected from the top of the Lower Magnesian Limestone. The map is largely based on borehole information with unpublished seismic interpretations by G A Kirby and J L Swallow for the north-east part of the district. The principal structures affecting the Permo-Triassic strata are named.

(Institute of Geological Sciences, 1973; British Geological Survey, 1983). The fault belt also affects the Permian, Triassic and Jurassic sequences, indicating reactivation in post Jurassic times.

## STRUCTURES AFFECTING THE POST-CARBONIFEROUS ROCKS

The Permian, Triassic and Jurassic rocks in the district dip eastwards at low angles (1 to 2°). Generally, the Permian rocks rest unconformably on a peneplained surface, except across the Harrogate Anticline where the Carboniferous rocks formed a topographical high. This high was overstepped by the Permian rocks, which thin out and are successively overlapped. The major structures affecting these post-Carboniferous rocks are mainly east-north-east-trending monoclines and fault zones forming the westerly extension of the Howardian Hills Fault Belt (Kent, 1980). Their continuity with this feature cannot be demonstrated in detail, however, because the lines of the faults cannot be traced through the intervening outcrop of homogeneous Sherwood Sandstone. From north to south these structures are: the Burton Leonard–Tollerton Monocline, the Ripley–Staveley Fault Belt, the Bilton–Scriven Monocline and the Rudfarlington–Forest Moor–Goldsborough Fault Belt. Additionally, there is the north-west-trending Ribston Hall Fault in the south of the district (Figure 20).

The Burton Leonard–Tollerton Monocline is an eastward-trending, northward-facing faulted monocline situated between grid lines 61 north and 65 north. It displaces the outcrop of the Permian rocks by several kilometres and brings the Mercia Mudstone Group and younger beds into the north-eastern corner of the district (Figure 3). The effective northward downthrow shown by the structure contours is between 150 m and 250 m (Figure 20). The upper hinge of the monocline is cut through by the Ripley–Staveley Fault Belt, which is responsible for the narrow graben preserving outliers of Lower Magnesian Limestone [294 599] exposed near Ripley, and of Sherwood Sandstone [325 605] proved by a borehole south of Brearton.

The Bilton–Scriven Monocline is an east–west, southwards-facing structure, marked by dips of up to 50°; it causes the regional north–south strike of the Permian rocks to swing east–west over a length of 5 km and is broadly coincident with the northern limb of the Harrogate Anticline. The monocline, along with the overstep of the Permian and Triassic sequences across the Harrogate Anticline, is responsible for the existence of the Bilton Hall outlier of Upper Marl and Sherwood Sandstone.

The Rudfarlington–Forest Moor–Goldsborough Fault Belt marks the southern edge of the Harrogate Anticline. It is a complex graben and horst structure parallel to the Ripley–Staveley Fault Belt. At its easternmost mapped extent, this structure coincides with a gravity high interpreted by Allsop (1985) as a probable horst of Dinantian strata caught in the fault zone.

The north-west-trending Ribston Hall Fault, south of Goldsborough, has a downthrow to the east and brings the Permian sequence against the Sherwood Sandstone Group. Minor faults of similar trend also occur along the same line in the vicinity of Burton Leonard. Other faults of this orientation have also been proved by geophysical surveys in the centre and east of the district.

Several eastward or east-north-east-trending faults affecting the Triassic and Jurassic rocks in the north-east of the district are inferred on limited borehole evidence, supplemented by resistivity and shallow seismic traverses carried out by the University of Leeds. Farther south, structures on these and other trends have been identified by recent (confidential) seismic investigations by British Coal and oil companies, and confirmed by British Coal drilling. Many of these structures become less intense westwards and only a small number affect the Permian outcrop.

Some of the structures affecting the Permian and younger rocks represent reactivation of pre-Permian lineaments, comparable to those in the Yorkshire coalfield to the south (Kent, 1974, fig. 4). Others appear to have formed during a period of north–south tension. Their age cannot be proved in this district, but evidence to the east and north-east (Hemingway and Riddler, 1982; Kent, 1974, 1980; Kirby and Swallow, 1987) suggests that they are pre-Upper Cretaceous, and may be related to the development of the Cleveland Basin in Jurassic times.

SEVEN

# Quaternary

About 85 per cent of the Harrogate district is mantled by Quaternary Drift deposits, the bulk of which are of glacial and periglacial origin. Most date from the late Devensian cold stage (Dimlington Stadial, approximately 18 000 to 14 000 years ago) (Rose, 1985; Bowen et al., 1986), but some older deposits may also be present, mainly in the south-west of the district. Postglacial (Flandrian) river deposits locally overprint the Devensian features.

The account that follows is in two parts. First, the history of Quaternary events in the district and its surrounding area is given in chronological sequence, so far as this can be inferred from the apparent spatial and stratigraphical relationships of the deposits; an underlying assumption is that since the glacier ice wasted progressively northwards, associated meltwater deposits started forming later in the north than in the south. Second, a description of the deposits, which is based on the categories of deposit shown on the drift edition of the published Harrogate sheet, and which is largely lithological in character, is given. The stratigraphical relationships between the mapped categories are shown in Figure 22.

## HISTORY OF QUATERNARY EVENTS

During the Devensian glaciation, ice encroached into the district down the Vale of York from the north. The main glacier was derived from the Lake District and crossed the Pennines by way of the Stainmore gap before splitting into two parts around the Cleveland Hills and North Yorkshire Moors. The western tongue extended southwards down the Vale of York, probably reaching the Isle of Axholme in north-western Lincolnshire (Gaunt, 1976) before quickly wasting back to positions of more prolonged still stand, when terminal moraines were formed at York and Escrick (Figure 21). Local valley glaciers entering the district from the Yorkshire Dales added to the Vale of York deposits along their western side. To the south-east of the district, North Sea ice moved across Holderness and into the Humber gap (Kent, 1980), effectively blocking the drainage from the Vale of York and impounding a large proglacial lake south of the York and Eskrick moraines.

As the Vale of York ice sheet stagnated, the ice front retreated intermittently and subordinate regressional moraines formed, belts of eskers were deposited, and short-lived glacial lakes developed in the Harrogate district north of the York moraine. The fluctuating advance and retreat of the Devensian ice front in the Vale of York resulted in a rapidly changing palaeogeography (Figures 24–27 and 29) and a complicated sequence of Drift deposits (Figure 22).

## Pre-Devensian or early Devensian

### HARROGATE TILL

The south-west of the district is mainly high ground, about half of which is covered by a dissected sheet of till. Both Tillotson (1932) and Edwards (1938) pointed out that this deposit, here distinguished as the Harrogate till, is markedly different from the till occurring east of the River Nidd (the Vale of York till). The Harrogate deposit is generally a heavy, slightly sandy clay, with large angular blocks, mainly of local sandstone and ganister; limestone is largely absent, possibly as a result of intense and protracted decalcification. Locally, the till becomes sandy over areas of sandstone bedrock. By contrast (Figure 26), east of the River Nidd and Cayton Gill (both of which are marginal glacial drainage channels), the Vale of York till contains abundant gravel, cobbles and boulders, mainly of Pennine origin but including numerous farther-travelled erratics (Harmer, 1928).

It is probable that the Harrogate till is relict from a pre-Devensian glaciation. Alternatively, it may have been deposited from late Devensian 'Dales' ice (Raistrick, 1932; Tillotson, 1932; Edwards, 1938), which encroached southwards to the outskirts of Leeds when the Vale of York ice was at its position of maximum advance (Institute of Geological Sciences, 1977). The main Vale of York till is firmly assigned to the late Devensian glaciation (Gaunt et al., 1971; Gaunt, 1981).

### PERIGLACIAL EFFECTS

Before being overridden by the Devensian ice sheet, the district was subjected to aeolian erosion in a periglacial environment. Sand-blasted, faceted and polished stones, or ventifacts, have been found preserved beneath the Devensian till at Aldborough, near Boroughbridge [4056 6611] (Gaunt, 1970) and near Allerton Mauleverer [416 580] (Gaunt, 1981; Institute of Geological Sciences, 1976). Abundant ventifacts were also found to the north of Whixley [451 592], but they were loose and could not be related stratigraphically to the till. South of the district, just outside the limits of the Devensian-glaciated area, Edwards (1936) also recorded numerous ventifacts. Some of these predate terrace gravels assigned to the York–Eskrick moraine interval of the Devensian glaciation.

Also peripheral to the main Devensian deposits, the formation of sculptured gritstone tors such as those at Plompton Rocks [356 537] has been attributed to periglacial action. Freeze-and-thaw associated with cambering, solifluction and aeolian erosion are thought to have been the main causative agents (Palmer and Radley, 1961). It is likely that the process of tor formation at Plompton Rocks continued throughout the Devensian

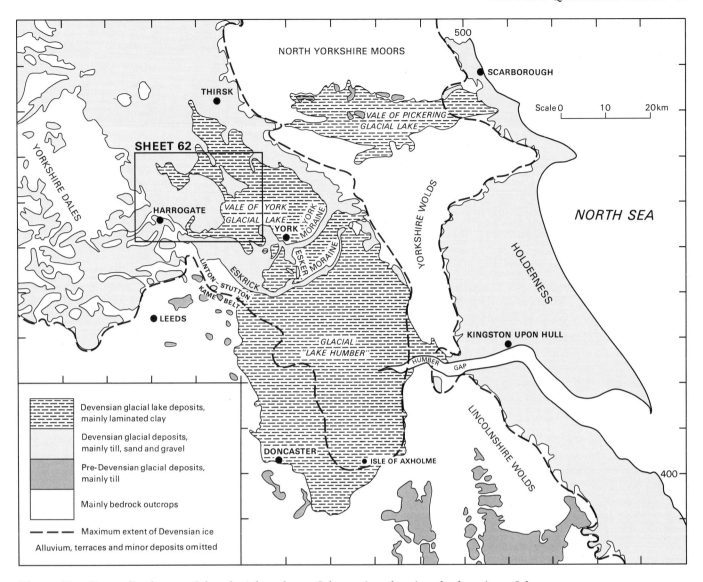

**Figure 21**   Generalised map of the glacial geology of the region showing the location of the Harrogate district.

glaciation, except for the short interval when they were covered by the ice at the glacial maximum. Meltwater flowing beneath the ice at around this time might have been responsible for the erosion of the channels which cut through Plompton Rocks and the surrounding area.

BURIED VALLEYS

The Pleistocene glaciations led to worldwide falls in sea level (Bowen, 1978; Gaunt et al., 1974; Gaunt, 1981), which in the Harrogate district caused the rivers to cut down to well below their present level. The rockhead contour map (Figure 23, and inset map on the Harrogate 1:50 000 sheet) reveals a pattern of buried valleys and shows that the bedrock surface over approximately one-quarter of the district lies below Ordnance Datum. Three major buried river valleys are identifiable, the proto-Ure, proto-Swale and proto-Nidd (Figures 23 and 24). The proto-Ure follows a similar course to the pres-

ent River Ure, but curves a little further south in the vicinity of Roecliffe [37 65]. Similarly, the proto-Swale follows essentially the same course as the present River Swale. The proto-Nidd, however, followed a course considerably different from that of the modern river, which occupies a glacial diversion course (Figures 23 and 24; and Johnson, 1974). From the west to near Ripley [285 595] it followed the present line of the river, but then headed in an easterly direction to near Farnham [345 605] before turning to the south-east and flowing for about 4 km to near Knaresborough [370 575]. From here it took an easterly then a south-easterly course, cutting through the Sherwood Sandstone escarpment where the present river does, at Kirk Hammerton [465 550]. Eastwards, it deviated again and headed to the south-east to near Rufforth [54 52]. The part of the proto-Nidd valley from Ripley to near Knaresborough is between one and two kilometres wide; eastwards from

**Figure 22**  Generalised relationships of the Quaternary deposits in the Harrogate district. Names of categories shown on the published 1:50 000 map are given in capitals.

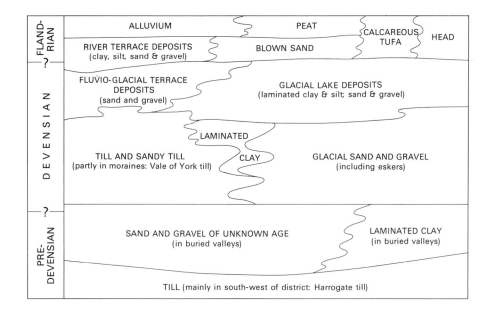

here, the valley broadens considerably as the soft Triassic sandstones are approached, its width suggesting that it was established for a considerable length of time before it was blocked by the Devensian ice sheet.

From near Farnham [345 615] to Staveley [360 625], the rock head contour map also shows a north-easterly trending buried valley connecting the proto-Nidd valley with the proto-Ure valley. This link is the Occaney gap, the floor of which slopes to the north-east; it is not clear when this channel was operative, but it is possible that the initial proto-Nidd headed to the south-east past Farnham, then later broke through the gap to head to the north-east past Staveley. This channel was reused during subsequent glacial deposition in the area. An alternative interpretation is that the Occaney gap was the initial river course, later blocked by the Devensian ice sheet, so that the Farnham buried valley represents only a glacial diversion. However, this interpretation appears unlikely because of the large size and width of the Farnham valley.

The buried valley system is partially filled with deposits classified on the map as Sand and Gravel of unknown age (in buried valleys); these include two distinct deposits of sand and gravel (Figures 24 and 25). Throughout most of the valley system the lower deposit is of probable fluvial origin (p.62.; Figure 24). In the south-western part of the district this is overlain by the Farnham fluvioglacial fan deposit and associated glacial lake deposits (p.62; Figure 25). All these deposits are concealed beneath the Devensian till.

**Late Devensian**

DEVENSIAN ADVANCE

As the Devensian ice sheet advanced towards its maximum extent, it diverted the proto-Nidd, Swale and Ure drainage towards the west and south (Kendall and Wroot, 1924), and impounded a glacial lake along its western margin (Figure 25). The overflow from the glacial lake, southwards into the Wharfe valley, was probably via the

Sike Beck channel, which is now concealed beneath the York moraine, or via the Kirk Deighton channel (Figure 25). Drainage from the ice sheet into this lake resulted in the formation of fluvioglacial outwash and glacial lake deposits, which partially fill the valley system. At Farnham, a proximal fan of sand and gravel was deposited, passing into progressively more distal deposits of sand, silt and clay to the south-east (Figure 25). These deposits were overridden by the further advance of the late Devensian ice and are now concealed beneath till in the buried valley system described above; the only exposures are in the Farnham gravel pits [35 60].

The maximum advance of the ice sheet was shortlived and south of the district left only deposits like those of the Linton–Stutton Kame-belt (Edwards et al., 1950), marginal channels, and fluvioglacial terrace deposits (Kendall and Wroot, 1924; Raistrick, 1932) to mark its edge. The line of maximum ice advance in the south of the Vale of York (Figure 21 and Institute of Geological Sciences, 1977) approximately lines up with the Nidd gorge at Knaresborough, and with the dividing line between the two types of till. It is probable that the glacial diversion drainage route of the Nidd was initiated at this stage. Following a period of ice-wasting a new ice front was established farther north, and the York and Escrick moraines were developed.

DEVENSIAN RETREAT

*Eskrick phase*

Following its maximum advance down the Vale of York, the ice front retreated to Eskrick. It was static for a considerable time and formed the well-developed lobate Eskrick terminal moraine (Figure 21; Kendall and Wroot, 1924; Edwards, 1938; Edwards et al., 1950). Deposits and features relating to this phase in the Harrogate district are shown in Figure 26. To the south of Knaresborough and west of Little Ribston, there is a thin covering of pebble and cobble till extending westwards to Crimple Beck; this appears to relate to the Eskrick position of the ice

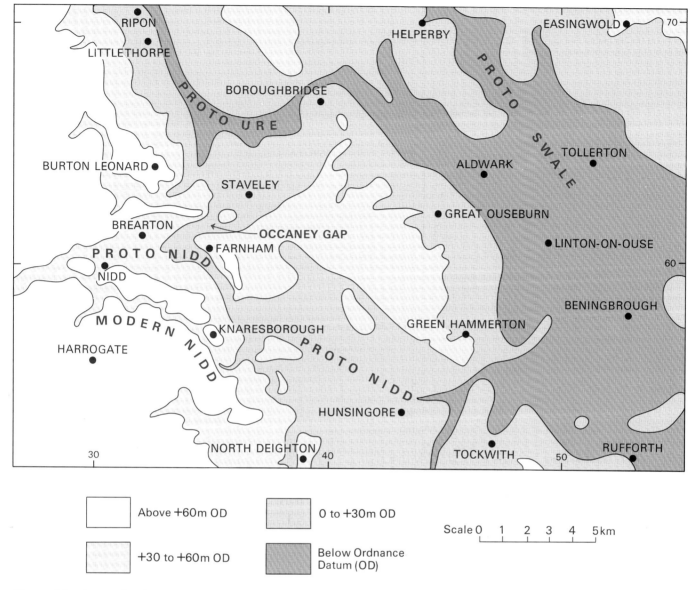

**Figure 23**  Generalised rockhead contours for the district.

front, but alternatively could relate to the slightly earlier maximum advance. The bedrock in the vicinity is cut by numerous humped (sensu Sugden and John, 1976, p.308) and incised channels, especially in the vicinity of Plompton Rocks [356 537]; these are interpreted as sub-ice drainage channels cut under hydrostatic pressure. That part of Crimple Beck between Oak View Farm [3445 5375] and Spofforth, just to the south of the district, may itself be a marginal drainage channel. It is possible that at this time the drainage diverted by the ice along the Knaresborough gorge was blocked by ice south-east of Knaresborough and spilled southwards via the Plompton channels and Crimple Beck. From here it drained southwards along the Kirk Deighton channel into the Wharfe valley near Wetherby. Other marginal drainage features appear to be associated with the York phase of the glaciation and are discussed below.

*York phase*

Yet another retreat, about 10 km north, established the terminal feature of the York moraine with its lobate form subparallel to the Eskrick moraine (Lewis, 1894; Kendall and Wroot 1924; Edwards, 1938; Edwards et al., 1950). The edge of this moraine just enters the district between North Deighton and Long Marston, where it turns to the north-west and forms a marginal moraine parallel to the Knaresborough gorge glacial drainage diversion (Figures 21 and 27). From Knaresborough northwards it appears that the marginal moraine is a composite feature formed by deposition from the ice responsible for the Eskrick, York and Flaxby moraines (Figure 27).

Diverted drainage associated with the edge of the ice-sheet, at both the York and Eskrick positions, was concentrated between the glacier and higher ground to the west. This led to the erosion of marginal channels (Fig-

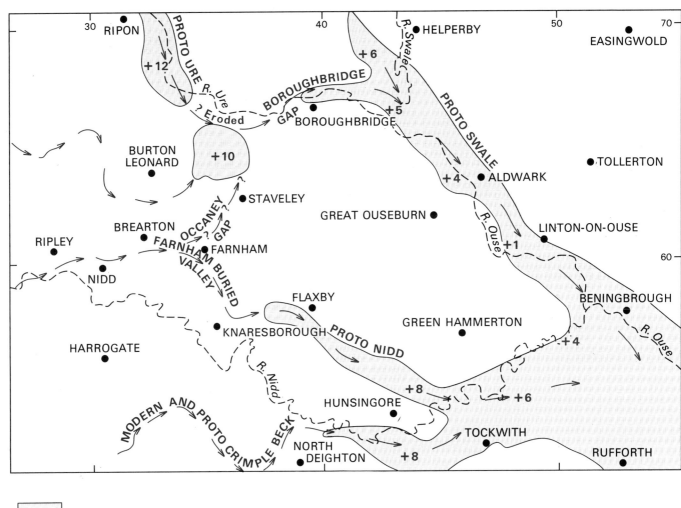

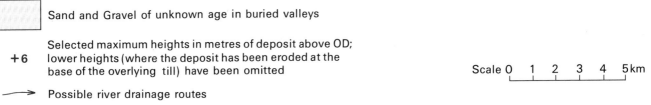

Sand and Gravel of unknown age in buried valleys

**+6**  Selected maximum heights in metres of deposit above OD; lower heights (where the deposit has been eroded at the base of the overlying till) have been omitted

Possible river drainage routes

Present river courses

Scale 0  1  2  3  4  5km

**Figure 24**  Distribution and maximum elevations of Sand and Gravel of unknown age (in buried valleys); possible river drainage routes are shown.

ure 26). Cayton Gill [286 628], a prominent valley 100 to 200 m wide and 4 km long, with very minor present-day drainage, was formed by water diverted southwards to join the Nidd. The Nidd was in turn diverted southwards to merge with water diverted eastwards by Wharfedale ice from the Washburn valley, along the deep channel running via John of Gaunt's Castle [220 546] and Oak Beck [292 559]. At the confluence of the Nidd and Oak Beck, a small glacial lake with its associated laminated clay deposits formed. This drained eastwards through the Knaresborough gorge, a rock-cut channel extending for about 8 km from near Nidd [30 59] to Knaresborough [36 56] (Figure 26). The drainage then ran south-

wards and joined up, via the Kirk Deighton channel, with that of the Wharfe valley, south-west of the York and Eskrick moraines (Edwards et al., 1950).

At the York, Eskrick and later phases, en- and supra-glacial drainage appears to have been concentrated in linear belts along which eskers developed. Between the York and Eskrick moraines a well-developed esker-like ridge runs south-south-east from York (Figure 21). This lines up with the Helperby–Aldwark esker that formed north of the Flaxby and Tollerton moraines described below (Figure 27). Northwards from here this esker can be traced for about 30 km across the Harrogate and Thirsk districts (Powell et al., in press); its total length is

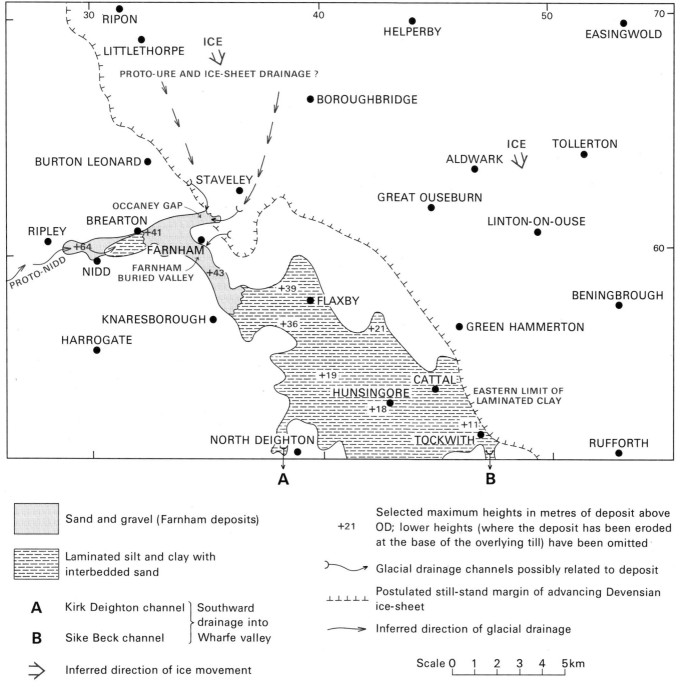

**Figure 25** Distribution and maximum elevation of Sand and Gravel of unknown age (with associated silts and clays) relating to the Occaney gap and Farnham buried valley. The postulated stillstand position of advancing ice is indicated.

at least 50 km. It appears that the main drainage routes from the ice sheet were long lived, but occasionally migrated slightly. Drainage spilling out on to the York moraine near the line of the Hunsingore esker was probably responsible for the deposition of a belt of fluvioglacial terrace deposits which extend southwards from Deighton Grange [410 516] (Figure 27). These join up with terrace deposits associated with the River Wharfe in the Leeds district to the south.

*Flaxby–Tollerton phase*

A further retreat of the ice front is documented by the presence of another belt of moraines about 10 km north of the York moraine, here named the Flaxby and Tollerton moraines (Figure 27). Each moraine has a lobate shape, the Tollerton moraine being offset northwards by about 3 km from the Flaxby moraine. The boundary between the two is marked by the Hunsingore esker. It is

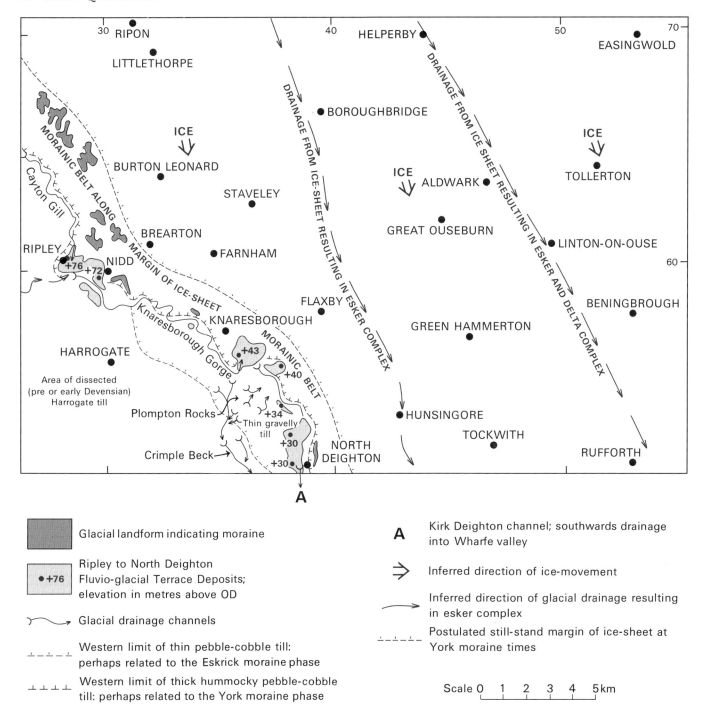

**Figure 26** Distribution of moraines, fluvioglacial terrace deposits and marginal drainage channels relating to the York and Eskrick moraine positions of the Devensian ice-sheet.

likely that the esker marks the dividing line between two streams of ice, possibly separated by the bedrock high of the Sherwood Sandstone escarpment. This line of weakness then became the focus for englacial drainage and the deposition of the Hunsingore esker, which also lines up with an esker belt in the Thirsk district to the north (Powell et al., in press). In the vicinity of Linton-on-Ouse, a transition from esker to delta is suggested by the presence of a broad fan of sand and gravel immediately

south of the Tollerton moraine. This is continuous with the Helperby–Aldwark esker complex (Figures 27, 28 and 29) and appears to have been deposited in a glacial lake impounded north of the York moraine.

*Final phases*

The closing phases of the Devensian glaciation saw the stagnation and wasting away of the ice sheet. The York and Eskrick moraines continued to act as a dam across

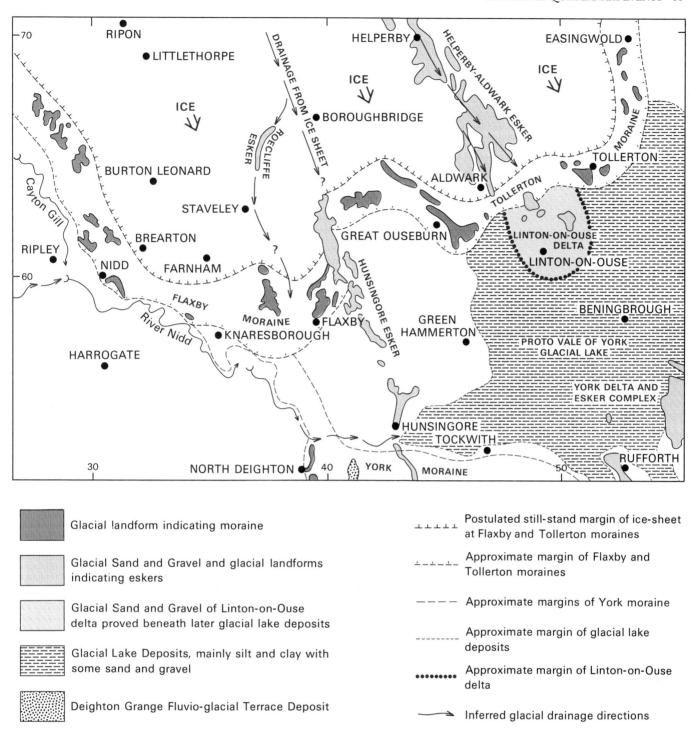

**Figure 27** Distribution of moraines, eskers, proglacial lakes and deltas relating to the Flaxby–Tollerton position of the Devensian ice sheet.

the Vale of York, and the extensive Vale of York glacial lake was impounded to the north of them (Figure 29).With the wasting of the ice sheet, much of the drainage reverted to near its preglacial course and the rivers Ure and Swale re-established themselves; the River Nidd was confined to its glacial diversion channel. All these rivers flowed into the Vale of York glacial lake; where they rivers entered the lake, they deposited most of the coarse material as Fluvioglacial Terrace Deposits (p.69), especially adjacent to the River Ure in the vicinity of Ripon and along the River Nidd near Hunsingore. These terraces are of sand and gravel proximally, and they grade distally into glacial lake deposits, first of sand, then of silt and clay. Eventually, the Devensian ice in the North Sea wasted away from the Humber gap and the glacial lakes both south and north of the York and Es-krick moraines drained away. The rivers established themselves across the lowest parts of the lake deposits and began to incise their present courses. In the periglacial environment, aeolian erosion of the unconsolidated glacial and postglacial sediments occurred and Blown Sand (p.72) was deposited as low dunes both on the flat lake deposits and on the rising ground to the

east of the glacial lake. In the York area, the age of these blown sand deposits probably spans the Devensian–Flandrian boundary with an age younger than 10 700 to 9950 years BP (Matthews, 1970). Deposits of Head (p.73) and the initiation of landslips may also date from around this interval.

### Postglacial to present day

The past 10 000 years or so, representing the Flandrian Stage of the Quaternary, have seen mainly erosional activity, with the incision of the rivers along their present courses. Numerous River Terrace Deposits (p.72) have been left in the process and a few of them are sandy or gravelly, especially in their lower parts. Many of the terraces grade into the Alluvium of the present floodplains (p.73) and it is difficult to work out any systematic chronological sequence for them. The present floodplains of the rivers are generally between 0.5 and 1.5 km wide, except where the rivers are confined in glacial diversion gorges, or where they cut through strong glacial features. The rivers are prone to flooding, but this has partly been prevented by the construction of artificial

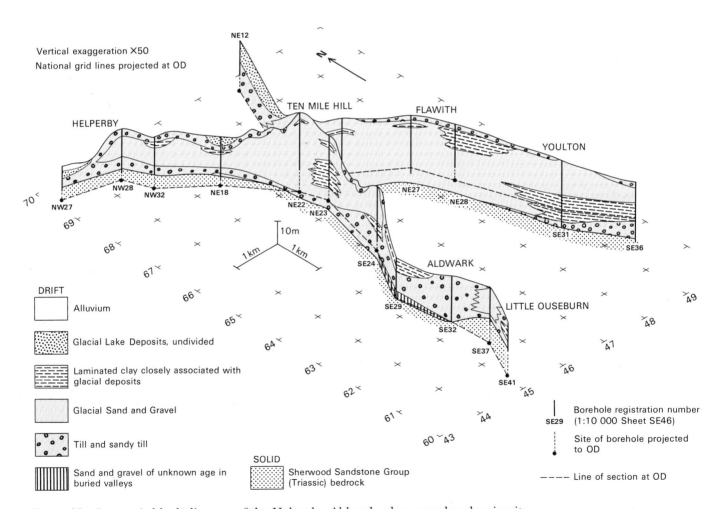

**Figure 28** Isometric block diagram of the Helperby–Aldwark esker complex showing its longitudinal and lateral variation; vertical exaggeration 50 times horizontal. Based on borehole information presented in detail by Stanczyszyn (1982).

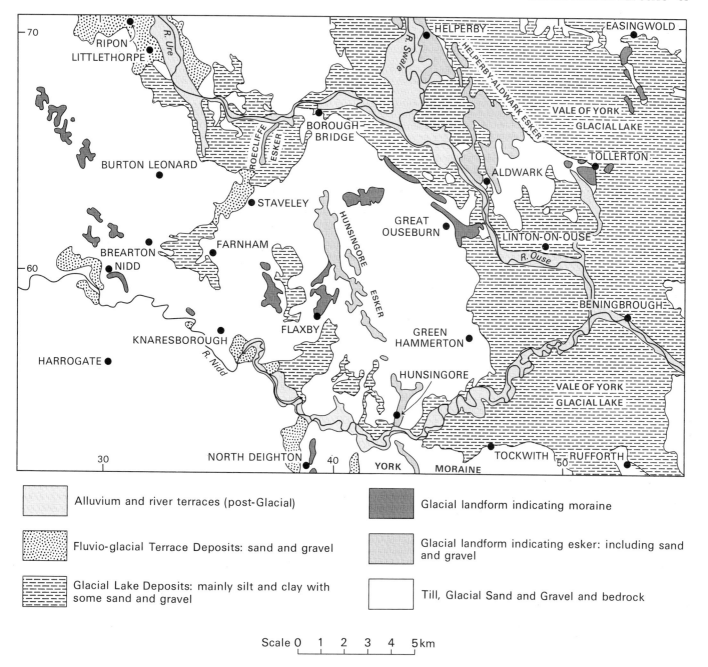

**Figure 29** Present day distribution of glacial deposits and the extent of the Vale of York glacial lake impounded north of the York moraine.

levees. This postglacial interval has also seen the development of Peat (p.74) in glacial kettle holes and in subsidence hollows caused by the underground dissolution of gypsum. Areas of calcareous tufa (p.73) related to spring activity also largely date from this period of time.

## DESCRIPTION OF DEPOSITS

Description of the deposits is arranged broadly according to the categories distinguished in the published 1:50 000 geological map. These categories are intended primarily to show the lithology of the deposits, with further subdivision, where possible, on the basis of morphology or stratigraphical position. Thus, several generations of sand and gravel (Sand and Gravel of unknown age in buried valleys; Glacial Sand and Gravel; Fluvioglacial Terrace Deposits) are identified, even though they may be lithologically similar. Some categories contain deposits of several ages, but cannot be subdivided on the map, with the result that the overall arrangement of the descriptive text cannot be strictly chronological; within each category, however, the deposits are described from oldest to youngest. A generalised view of the stratigraphical relationships between the mapped categories is shown in Figure 22.

## Sand and Gravel of unknown age (in buried valleys)

Two distinct deposits of sand and gravel underlying the Devensian till in the district have been placed in the 'unknown age' category (Figure 22) on the published map. They are the Ripon to Rufforth deposit (Figure 24), and a deposit around Farnham (Figure 25). Although they are included in the same category the two deposits appear to have different origins and ages. However, they have not yielded any palaeontological or radiocarbon evidence and their precise relationships are speculative. The Ripon to Rufforth deposit predates the late Devensian till, but is otherwise undated, although deposits of Ipswichian age have been recorded from similar valleys in the southern Vale of York (Gaunt et al., 1974). The Farnham deposit also predates the till, but has a restricted extent and appears to be related to a stillstand during the advance of the Devensian ice sheet.

### The Ripon to Rufforth deposit

Numerous mineral assessment boreholes, sited in the northern, central and eastern parts of the district, have proved sand and gravel partially filling the buried valley system described above (Figures 23 and 24) (Abraham, 1981; Stanczyszyn, 1982; Giles, 1981). South of Ripon, along the course of the proto-Ure to near Staveley, sand and gravel is up to 9 m thick, with subordinate laminated clay. The morphology of the deposit is difficult to assess, due to erosion by the Devensian ice sheet and foundering caused by gypsum dissolution (see chapter eight), but the top of the deposit generally lies between about 14 and 9 m above OD, sloping gently southwards from Ripon. The deposit also occurs at the north-east end of the Occaney gap (Figure 24) and may include some deposition here from the proto-Nidd. Marginal to the sand and gravel, laminated clay, up to 9 m thick, also occurs; this may be partly a glacial lake deposit related to the blocking of the Boroughbridge gap by the advancing ice sheet. From Roecliffe north-eastwards through the Boroughbridge gap, both the sand and gravel and the associated laminated clay are missing, presumably as a result of erosion by the proto-Ure and subsequently by Devensian ice. The sand and gravel deposit comprises about 9 per cent silt and clay, 53 per cent sand and 38 per cent gravel. The sand fraction is mainly coarse and medium quartz with some local lithic grains, and the gravel component is dominantly of Carboniferous sandstone and limestone, with Permian limestone and small amounts of mudstone, chert and quartzite; precise details are given by Abraham (1981).

East of the Boroughbridge gap, the sand and gravel forms a spread approximately following the line of the present River Ure. The deposit has an upper surface at around 6 m above OD, sloping to the south-east, and

comprises gravelly sand up to 8 m thick with an overall mean thickness of 4.6 m (Stanczyszyn, 1982). The ratio of sand to gravel is similar to that noted above and the gravel content is similarly dominated by Carboniferous sandstone and limestone with lesser amounts of Permian limestone. Small amounts of Triassic sandstone and ironstone reflect the local bedrock and the closer proximity of the Jurassic ironstone sequences respectively; full details are given by Stanczyszyn (1982).

In the southern half of the district the sand and gravel deposit, occupies the proto-Nidd valley south-eastwards from about Flaxby, and the proto-Crimple valley eastwards from Little Ribston. It then extends eastwards into the Vale of York, joining the deposits of the proto-Ure and proto-Swale in the south-east corner of the district, but details here are scant. At its western limits, the deposit has a surface elevation of about 10 m above OD, falling to just below OD in the east. In the proto-Nidd valley, its upper surface, west of the Sherwood Sandstone escarpment, is difficult to identify due to the presence of the overlying Farnham deposit (see below) and its associated laminated clays. East of the escarpment, the upper surface of the deposit has suffered erosion at the base of the Devensian ice-sheet and can be as low as 4 m below OD. The more distal nature of the deposit here is reflected in the composition, which includes about 9 per cent silt and clay, 66 per cent sand and only 25 per cent gravel. The sand fraction is mainly fine and medium grained, and hard Carboniferous sandstones and limestones are the dominant gravel clasts; subordinate amounts of quartz, chert, ironstone and shale are also present in varying quantities; full details are given by Giles (1981).

### Farnham deposits and related silts

As the Devensian ice sheet advanced southwards, it blocked the rivers entering the Vale of York from the west and diverted them along its edge (Figure 25). The drainage of the diverted proto-Ure, combined with the south-flowing drainage from the ice sheet, spilled out westwards through the now partially buried Occaney gap [352 619]. Here it joined the proto-Nidd, which at that stage must have flowed to the north of Knaresborough through the now buried valley of Brearton and Farnham (Figure 24). Here, two fans of sand and gravel were deposited in the preglacial Nidd valley (Figure 25). The larger deposit is closest to source at Occaney and Farnham, where a probable glacial drainage channel incised in the escarpment runs down to Farnham village. The sand and gravel here thins westwards towards Brearton and southwards towards Knaresborough, passing laterally into laminated clay deposits of a glacial lake in the southeast. The upper surface of these deposits lies at around 40 m above OD in the vicinity of Farnham gravel pit [345 600]; it falls northwards to about 28 m above OD east of Brearton [335 607], but appears to rise again westwards past Brearton, where it interfingers with a smaller fan deposit of sand, gravel and laminated clay deposited from the proto-Nidd. To the south-east, the maximum surface elevation of the sand and gravel, passing distally into silt and clay, falls gently to around 11 m above OD in the vicinity of Cattal [445 546]. In places, the surface of these deposits is much lower and has an uneven nature. This may be attributed to erosion at the base of the Devensian ice sheet, which overrode them and deposited the overlying till.

The sand and gravel deposit is exposed beneath Devensian till in the Farnham sand and gravel pits [3475 6032], where it ranges from boulder gravel to fine sand; the clasts are mainly of Carboniferous sandstone and limestone, but Permian limestone is locally abundant (Abraham, 1981; Dundas, 1981). Sedimentary structures, formerly exposed at Farnham quarries [353 590], include braided channel deposits, probably formed in an alluvial fan or sandur environment.

Around Farnham, the deposit is completely composed of sand and gravel, but south-east from near Knaresborough [358 583] laminated clay interdigitates with this coarser material. Distally, towards Cattal [445 546] in the east, the deposit thins, the clay and fine sand proportions increase, and the percentage of gravel decreases (Abraham, 1981; Dundas, 1981; Giles, 1981). When these deposits were laid down, ice was probably blocking the easterly channel through the Sherwood Sandstone escarpment, forcing the drainage southward towards the Wharfe valley through the now largely drift-filled Sike Beck channel, or possibly through the Kirk Deighton channel (Figure 25).

## Till and Sandy Till

Within the district two distict areas of till are present. The possible age relationships of the two are discussed above (p.52 and Figure 22).

### Harrogate till

The till in the south-west is generally a heavy, slightly sandy clay, with large angular blocks, mainly of local sandstone and ganister; limestone is largely absent, possibly as a result of intense and protracted decalcification. This till mainly caps the higher ground and has been largely eroded from the valley sides. It contrasts markedly with the Vale of York till, which blankets the low ground to the east of the glacial drainage channels at the probable edge of the main Devensian glacial deposits. Sections in the heavy Harrogate till are scarce and show very little lithological variation. Where there is sandstone bedrock, the till tends to be sandier, but it rarely approaches a sand deposit; areas of glacial sand and gravel within it are minimal.

### Vale of York till

By contrast, the Vale of York till over the remaining area is much more sandy and includes dominantly rounded and subrounded pebbles and cobbles. These erratics are farther travelled than those in the Harrogate till, though local bedrock also forms an important part of the stone content. The till throughout this area is largely associated with deposits classified as Glacial Sand and Gravel (p.65) and has a fairly 'fresh' topography, which has suffered little from erosional degradation. The topo-

graphical expression of the glacial moraines and eskers with a few drumlins is still well preserved, allowing their palaeogeography to be reconstucted.

The York and Eskrick moraines, just south of the Harrogate district, represent stillstands of the Devensian ice sheet (Figures 21 and 27). The Eskrick moraine lies mainly to the south of the district, but its lateral margin probably entered the area around North Deighton [391 515]. It may, however, have suffered later erosion and only a spread of thin till is now present west of the village (Figure 26). Alternatively, it is possible that this thin till may relate to the maximum advance of the Devensian ice sheet as described above (p.54). The northern slope of the York moraine forms a ridge of hummocky ground extending from where it enters the district, south of Long Marston [500 513], westwards to North Deighton [391 515] (Figure 27). From there, in a north-westerly direction through Knaresborough, patches of marginal moraine are aligned parallel to the glacial diversion channels of North Deighton, the River Nidd and Cayton Gill, which ran outside the moraine belt. Inside the moraine belt, pebble- and cobble-rich Devensian till is present over most of the ground to the north and east. North of the York and Eskrick moraines, remnants of the Flaxby and Tollerton moraines, extending from Knaresborough to Easingwold, mark a further stillstand or retreat of the ice. The eskers which intersect the lobate form of this belt suggest that several contemporaneous ice streams may have contributed to its formation (Figure 27).

Most of the Harrogate district, especially north and east of the marginal moraines and channels, is blanketed by till and sandy till, much of which is concealed beneath later glacial lake deposits. The till and sandy till commonly grade into each other and no geological lines could be drawn between them; their distribution therefore is indicated only by symbols on the published map. The till represents both the bedload of the late Devensian ice sheet (lodgement till) and the suspended and supraglacial load (ablation till). It has an average thickness up to about 20 m, but may be considerably thicker in moraines and buried valleys. It produces an undulating topography, hummocky in places, with abundant enclosed hollows, or kettle holes, in which later deposits of peat and clay have accumulated — a landscape associated typically with the melting of a stagnant ice sheet. Around Spa Well [514 695] and Gallows Hill [503 698], sandy clay and clayey till occur in topographical depressions surrounded by younger lacustrine deposits at higher topographic levels. This suggests that when the surrounding lacustrine deposits were formed, the till here was either upstanding and supported by buried ice or protected from sedimentation by surface masses of stagnant ice. The ice subsequently melted causing the till to be exposed at the bottom of depressions in the lacustrine deposits.

The lithology of the till is closely related to that of the bedrock in the near vicinity. In the west, overlying Permian rocks, the till is generally brown and red-brown sandy clay, with local Permian limestone frag-

**Plate 8** Sandy till on gravelly till at Grafton Gravel Quarry [4200 6310].

The upper 0.8 m of the face is sandy till; this consists of a matrix of poorly sorted silty sand supporting predominantly Carboniferous sandstone pebbles, cobbles and boulders. The reddened nature and lack of limestone clasts in the upper 0.8 m is due to decalcification of the till by the downward percolation of meteoric water. The underlying gravelly till has a coarse-grained sandy matrix with pebbles, cobbles and boulders of Carboniferous limestone and sandstone (L 1798).

ments, as well as numerous erratic pebbles and boulders of farther-travelled Carboniferous sandstone and limestone. In the central belt, the dominant sandy till, largely derived from the Sherwood Sandstone bedrock, mainly comprises fine-grained, red-brown, clayey or silty sand containing some Sherwood Sandstone fragments, but more abundant subrounded pebbles and boulders of Carboniferous sandstone and limestone (Plate 8) are also present. In the east of the district, over the Mercia Mudstone and younger rocks, the till and sandy till comprise reddish brown to dark brown, fine-grained, clayey sand and sandy clay, with Carboniferous sandstone and limestone erratics, and numerous Jurassic fragments, including ironstone. Exotic erratics occur mainly to the east of the Sherwood Sandstone escarpment, suggesting that ice flows from two separate sources abutted here, the junction being marked also by a line of eskers. To the west of the escarpment, the

till contains mainly Carboniferous sandstone and limestone with some Permian limestone, suggesting a dominantly Pennine source area but, east of the escarpment the erratics also include basic igneous rocks and fine-grained acid volcanic rocks. Many of these rock types have been attributed to a Lake District source (Harmer, 1928); the suite includes boulders of the distinctive Shap granite. However, a Shap granite boulder was also recovered from near Deighton Grange [4174 5143] (see details below), which lies west of this line, so that some mixing of the component ice flows, or erosion of earlier deposits, can be inferred.

In many places, the upper 0.3 to 1.0 m of the till is decalcified and only the larger limestone clasts remain; these have rough, pitted and weathered surfaces. The decalcification is most intense in the higher, well-drained ground and it alters the upper part of the till to a red-brown colour; this is seen at Monkton Moor Quarry [3076 6518], where the upper reddened layer is 0.3 to 0.6 m thick, and at Grafton Gravel Quarry [4200 6310] (Plate 8) where it is 0.8 m thick. In Knaresborough railway cutting [3638 5886], the top 0.3 to 0.5 m of the till is similarly decalcified. In the lower poorly drained ground, and in places where the till has a clay matrix, the decalcification is less apparent; in these sections the erratics commonly show well-developed glacial striae.

The moraine deposits consist mainly of till, but glacial sand and gravel are abundant locally; subordinate areas of laminated clay also occur. The morainic features are elevated above the surrounding till plain and the deposits within them are commonly 30 to 40 m thick. Their distribution is shown in Figures 26, 27 and 29. The moraines generally have steep margins, which in some cases may be interpreted as ice contact features. A probable ice contact feature has been mapped for 1.5 km to the west of Grafton [415 633]; the slope dips at about 20° and falls from about 23 m to about 12 m westwards. Another probable ice contact slope forms the northern flank of the marked ridge of till which runs north-east from Coneythorpe [394 594] for about 1.5 km. In the west of the district, prominent ridges of till with steep eastern margins at How Hill [2767 6700] and near Ingerthorpe [284 662 and 288 657] form part of the marginal moraine that parallels Cayton Gill. Farther south, the strong moraine-ridge running northwards from North Deighton [393 518] similarly has a steep eastern side and marks the edge of the thick till that covers most of the Vale of York. The structures within the moraines are commonly complex, with interfingering lithologies and isolated pods of different materials. Distortion of the deposits is attributed partly to ice-pushing and partly to collapse of the deposits during melting of the ice.

## Details

Within the district, large sections through the till are uncommon, although abundant details were obtained from boreholes drilled for sand and gravel assessment. For these, reference should be made to Abraham (1981), Dundas (1981), Giles (1981), Morigi and James (1984) and Stanczyszyn (1982).

At Monkton Moor Quarry [3076 6518], the till overlying the Lower Magnesian Limestone is 1 to 1.5 m thick. The upper 0.4 m of the till is red-brown and decalcified with no Permian limestone fragments; below, the till is grey-brown with Permian limestone fragments in the process of disintegration. The erratic content is mainly Carboniferous sandstone, but Carboniferous limestone is also prominent.

Till is exposed at the sand and gravel quarries [3480 6017; 3444 6013; 3398 6004] near Farnham, where it overlies sand and gravel in a buried valley (see above). The till hereabouts is 2 to 4 m thick and comprises red-brown silty clay with rounded to angular pebbles, cobbles and boulders, up to 0.8 m in diameter, of Carboniferous sandstone, limestone and chert, and Permian limestone.

In the Knaresborough railway cutting [3638 5886], which was being filled with refuse at the time of the survey, the section in the till overlying the Lower Magnesian Limestone showed considerable lateral variation, but showed the following approximate sequence:

| | *Thickness* m |
|---|---|
| Topsoil with plant roots | 0.2–0.4 |
| Clayey sand, brown, with numerous granules to boulders of dominantly Carboniferous sandstone; limestone clasts uncommon to absent | 0.5–0.7 |
| Sand, red-brown, clayey, fine- to medium-grained, with sporadic pebbles | 0.2–0.3 |
| Gravel, composed of abundant pebbles and cobbles dominantly of Carboniferous sandstone, with up to 40 per cent of Carboniferous limestone, plus sporadic chert, Magnesian Limestone and ganister in a slightly clayey sand matrix | 0.6–0.7 |
| Sand, calcareous, brown, becoming reddish brown towards the top and containing abundant angular fragments of Magnesian Limestone and streaks of weathered Magnesian Limestone in the bottom 0.15 m | 0.5–0.6 |
| Weathered top of the Lower Magnesian Limestone, which on the south side of the cutting showed contorted clayey laminae that possibly represent cryoturbation structures | 0.2–0.5 |

This section shows decalcification of the upper part of the sequence.

Near Deighton Grange [4174 5143], a ditch section proved a low buried mound of till beneath later clay and peat. The section, about 2.5 m deep, showed brown and grey till with abundant erratics up to boulder size in a matrix which varied in patches from very sandy clay to slightly sandy clay. The erratic content was dominantly hard, white and grey sandstone with 10 to 15 per cent of grey Carboniferous limestone and a few chert clasts. The section also yielded a boulder of Shap Granite, 0.25 m in diameter.

Grafton Gravel Quarry [4207 6295] (Plate 9) showed complex relationships between the glacial sand and gravel, the overlying gravelly till and the sandy till at the surface. Nearby in the same gravel pit [4200 6310], the section through the till (Plate 8) showed an upper 0.8 m of reddish brown sandy till with pebbles, cobbles and boulders dominantly of Carboniferous sandstone. The underlying gravelly till had a coarse-grained sandy matrix with pebbles, cobbles and boulders of Carboniferous sandstone and limestone; the difference between the two till layers may be explained by superficial decalcification.

Road widening [5249 6991] near Easingwold police station exposed up to 3.5 m of clay with pebbles from which an erratic block of shelly ferruginous limestone, over 1.3 m long and of probable Jurassic age, was removed. Erratics of Jurassic ironstone are also common in the Vale of York as far west as the Sherwood Sandstone escarpment; numerous clasts of ironstone were found in the vicinity of Tockwith [470 520].

## Glacial Sand and Gravel

Deposits classified on the published map as Glacial Sand and Gravel are widespread in bodies ranging from small lenses to spreads up to several square kilometres in extent. Two sinuous elongate eskers have been mapped traversing the district and, apart from the accumulations in buried valleys (see above), comprise the main sand and gravel deposits.

The bulk of the deposits is fairly clean sand and gravel with a sandy matrix, although gradations from silt to

**Plate 9**  Distorted Glacial Sand and Gravel at Grafton Gravel Quarry [4207 6295].
The contorted fold in the centre of the photograph is composed of interfingering sands and gravels.
These waterlaid deposits are overlain by sandy on gravelly till. Deformation was probably
contemporaneous, a result of ice-pushing and the melting out of buried ice. (Spade is 0.8 m from handle
to top of blade.) (L 1795).

boulder gravel are present. However, some of the gravelly deposits, mainly those associated with the moraines, consist of unstratified cobbles and boulders with a small amount of clay matrix similar to that of the local till. The lithologies of the gravel clasts are similar to those found in the tills; thus these deposits were clearly also largely derived from the melting ice sheet. The clasts are mainly of Carboniferous sandstone, but in places Carboniferous limestone is abundant, forming up to 30 per cent of the stone content, and Magnesian Limestone, which is locally, widespread on the Permian bedrock locally forms up to 15 per cent of the gravel. Full details of the gradings and lithology of the sand and gravel deposits are given by Abraham (1981, Dundas (1981), Giles (1981), Morigi and James (1984) and Stanczyszyn (1982).

Two main south-south-easterly trending belts of sand and gravel occur (Figures 27 and 29) as irregular, anastamosing ridges with sharp marginal features and elevations of 10 to 20 m above the surrounding area. They are

the Helperby–Aldwark and the Hunsingore eskers, which are shown edged in pink on the Drift edition of the published geological map. Within them, the sand and gravel is partly concealed beneath a cover of till and in places protrudes as ridges through the later glacial lake sediments (Figures 28 and 29).

Between Helperby [440 700] and Aldwark [474 633] the sand and gravel forms part of the anastomosing Helperby–Aldwark esker (Figure 29). This feature forms part of an esker belt which extends for about 50 km from south of York to the northern edge of the Thirsk district (Powell et al., in press). In both this and the Hunsingore esker, the sands and gravels exhibit sedimentary structures such as cross-bedding, ripple marks, washouts, channels, slumps and load-casts (Plate 10); lateral passage into laminated clay, both by gradation and interdigitation, is common (Figure 28). The morphology and sedimentology of these deposits suggest deposition in a glaciofluvial environment in valleys upon the ice sheet, or possibly in tunnels under or within the ice (Banerjee and McDonald, 1975). Subsequent melting of the ice has largely inverted the topography, leaving the sand and gravel as upstanding ridges or eskers. Removal of the ice support led to synsedimentary deformation, such as slumping, minor faulting and folding or tilting.

South of the Helperby–Aldwark esker, the sand and gravel forms subdued, partially concealed spreads around Linton on Ouse airfield. This change to broad spreads occurs where the esker complex intersects the Tollerton moraine. The Linton on Ouse deposits are probably part of a delta system which formed where meltwaters issued from the front of the Devensian ice sheet while it was stationary at the Tollerton moraine position, and which spilled into a glacial lake impounded north of the York and Eskrick moraines. The Helperby–Aldwark–Linton on Ouse belt of sand and gravel has been proved beneath later glaciolacustrine deposits in boreholes south-eastwards as far as Beningbrough. It is aligned with spreads of sand and gravel at York and with a well-developed esker-like ridge running south-south-east from York between the York and Eskrick moraines (Figure 21). It appears that, as the ice sheet retreated, the main position of the drainage route along it remained fairly constant (Figure 26), resulting in the progressive formation of this linear sand and gravel belt through several retreat phases (Figures 26, 27 and 29).

In the centre of the district, a second esker complex, the Hunsingore esker (Figure 27), stretches southwards from near Brooms House [399 633] to Cowthorpe and Lingcroft [437 512], a distance of about 13 km. Abundant sand and gravel is exposed in the field brash, but much of the anastamosing ridge system is concealed by till. The esker deposit across Roecliffe Moor [375 658] may be part of this complex, but it is largely concealed by later glacial lake deposits.

## Details

### The Helperby–Aldwark esker

The Helperby–Aldwark esker can be traced for about 30 km north-north-west from the Tollerton–Flaxby moraine. The deposits of the esker were exposed in a gravel pit at Ten mile Hill [4658 6714], where a section showed sandy till overlying glacial sand and gravel. The following sequence was recorded:

| | Thickness<br>m |
|---|---|
| Clayey sand with pebbles, cobbles and boulders (some ice-scratched) and with lenses of clay with pebbles and fine to coarse, bedded gravel. The erratics are mainly Carboniferous sandstone, but Carboniferous limestone, Permian limestone and Sherwood Sandstone clasts are also present; one boulder of probable volcanic rock was seen. | 0–1.50 |
| Sand, red-brown, fine-grained, well-washed, with beds and laminae of brown silt and clay. Small-scale climbing ripples, in sets 0.02 to 0.09 m high, are widespread in the fine-grained sand. Some load-casting of the sand into the silt is present. | 4.50–6.00 |

The upper sandy till unit is discontinuous and interfingers with the fine-grained sand, which in its upper part passes laterally into gravel and sand displaying channel structures. The clay laminae in the lower part of the sand show contortions indicating sediment collapse, probably due to the removal of ice support.

A disused sand and gravel pit [4747 6615] between Tholthorpe and Flawith, also in the Helperby–Aldwark esker, showed complex relationships between the lithologies. The lithological units could only be traced for a maximum distance of 25 m, and many of them for only a few metres. The units displayed contorted contacts and had obviously undergone slumping, probably as a result of the removal of ice support. At the western side of the pit the following section was recorded:

| | Thickness<br>m |
|---|---|
| Sandy soil, dark brown, with a few pebbles | 0.30 |
| Sand, yellow-brown, fine-grained, generally well-sorted, but with a few pebbles | 0.10 |
| Sand, grey-brown, slightly clayey | 0.10 |
| Clayey sand and sand with pebbles and cobbles, gravelly in places (flow till?) | 0.50–1.30 |
| Clay, purple-brown and 'waxy', with sporadic patches of sandy clay and sand, and sporadic pebbles | 0–1.00 |
| Sand, red-brown, fine-grained, well-sorted, with laminae of silt and clay which pick out complex contortions and folds in the sand and clay | 1.40 |

Extending north-eastwards for 35 m from this section, the face was only 2 to 3 m high and showed more continuous beds with less slumping. The section here comprised a surface layer of dark brown sandy soil 0.25 m thick, over a sandy till layer composed of pale brown sand with a few pebbles and lenses of sand and gravel 0.5 to 1.0 m thick. This was underlain by a lenticular gravelly till layer of ill-sorted clayey gravel up to 0.6 m thick. The basal part of the section comprised fine- to coarse-grained sandy gravel 0.5 to 1.6 m thick, with lenses of fine-grained, clean, red-brown sand up to 0.75 m thick; channel structures and cross-lamination were the main sedimentary structures in the deposit.

### The Hunsingore esker

The Hunsingore esker can be traced for about 13 km in a south-south-east direction across the centre of the district from Brooms House [339 633] to Lingcroft [437 512] (Figure 27). At its northern end, where it intersects the Flaxby and Tollerton moraines, its morphology is fairly subdued. South of Claro House, it forms a train of elongate hills up to 0.5 km wide and 2 to 3 km long. These hills are generally steep on both their western and eastern sides, and are composed mainly of sand and gravel, but have fairly extensive coverings of thin till along their flanks. At Allerton Grange, a mineral assessment borehole [4167 5751] proved 10.2 m of clean sand and gravel before being abandoned against a boulder (Giles, 1981). Farther south, another borehole of the same survey at Nursery Hill Cottage [4300 5436] proved the following sequence, which illustrates the complexity of the esker deposits:

**Plate 10**   Glacial Sand and Gravel at the old sand and gravel pits [c. 4760 6620] between Tholthorpe and Flawith.

This temporary section showed approximately 1.5 m of well-washed, cross-stratified, red-brown, fine-grained sand with lenses of pebble and cobble gravel. The deposit comprises part of an esker complex (L 3079).

|  | Thickness m | Depth m |
|---|---|---|
| Soil | 0.5 | 0.5 |
| Sandy gravel, of granules, pebbles and cobbles, mainly of sandstone and limestone in a clayey sand matrix of mainly fine-grained quartz sand | 0.5 | 0.5 |
| Clay, sandy, with sporadic thin beds of sand and scattered pebbles (till) | 3.2 | 3.7 |
| Sand, fine-grained and clayey, with sporadic thin beds of laminated clay | 15.1 | 19.1 |
| Clay, sandy and pebbly (till) | 5.9 | 25.0 |

*The Roecliffe esker*

The ridge of glacial sand and gravel across Roecliffe Moor has an esker-like form (Figure 29). It extends from Roecliffe [377 661] in a south-westerly direction and then turns towards the south. The ridge is nearly 1.9 km long and stands 2 to 4 m above the surrounding flat of glacial lake deposits. In the north, a borehole [3768 6587] proved 24 m of sand and gravel down to bedrock, but the lower part may lie in a buried valley. Halfway along the ridge, a mineral assessment borehole [3723 6511] (Abraham, 1981) proved mainly uniform clayey sand with sporadic pebbles to a depth of 14.1 m, resting on 5.6 m of till, 1.2 m of sand and gravel (in a buried valley) and Sherwood Sandstone bedrock. The ridge starts abruptly in the north and dies out to the south in line with the Occaney gap where it may have been removed by later erosion associated with the fluvio-glacial terrace deposits. It may be related to englacial drainage (Figure 27) associated with a string of small sand and gravel deposits which extends southwards through Staveley, Ferrensby and Goldsborough to Ribston Hall [393 541]. Alternatively, it may be a spur from the main Hunsingore esker.

## Laminated clay closely associated with glacial deposits

Laminated clays, which occur interbedded both with the glacial sand and gravel and with the tills, crop out locally and were proved by soil augering around Great Ouse-burn [457 616], Little Ouseburn [444 610] and Golds-borough [378 564]. The surface outcop area of these deposits is very small, but they have also been proved below surface in numerous mineral assessment boreholes. Some of the deposits consist of finely interlaminated clay and silt, and were probably deposited in a glaciofluvial environment by glacial meltwater where the water flow was slack. In places, the glacial sand and gravel of the Helperby–Aldwark esker complex also pass laterally into laminated clay, which is extensive at the southern (distal) end of the esker (Figure 28). In a few places, mainly within the till deposits, the laminated clays are very stiff and compressed, possibly as a result of shearing and compaction beneath a considerable cover of ice.

## Fluvioglacial Terrace Deposits

Sands and gravels classified as Fluvioglacial Terrace Deposits in the Harrogate district are not subdivided on the published map, but three distinct sets are recognised in this account. The first set, between Ripley and North Deighton, was formed where marginal drainage, diverted by ice related to the Eskrick and York moraines (see above), crossed pre-existing valleys and low ground. The second set was deposited, in the vicinity of Staveley and Deighton Grange, during a later stillstand of the ice sheet. The third set was deposited around Ripon and is proximal to a suite of glacial lake deposits. Full details of gradings and lithology of these deposits are given by Abraham (1981), Dundas (1981), and Giles (1981).

### The Ripley to North Deighton deposits

The first set of fluvioglacial terrace deposits was formed where marginal channels, peripheral to the ice sheet, crossed pre-existing valleys and low ground (Figure 26). The terraces lie in isolated patches along the present course of the River Nidd, mainly near Ripley [295 603], east of Knaresborough gorge [366 563], and from Little Ribston [385 530] to the north end of the Kirk Deighton overflow channel. They have an elevation of about 76 m above OD where they spill from the Cayton Gill channel [2884 6132], drop to 43 m above OD where they exit from Knaresborough gorge, and reduce in height to 30 m above OD where they enter the Kirk Deighton glacial drainage channel. The deposits are generally composed of very clayey sand and gravel, with included pebbles and cobbles dominantly of Carboniferous sandstone and limestone, and subordinate Permian limestone. In places, the deposits pass into laminated clay, which becomes more extensive southwards towards the Kirk Deighton channel.

### The Staveley and Deighton Grange deposits

The second set of fluvioglacial terrace deposits is worked in a quarry at Staveley [360 626] (Figure 29). Here, the deposits form a gently sloping fan, spreading out from a rock-cut channel [347 628] at Copgrove. The sand and gravel terrace, with an elevation of about 38 m above OD, partially chokes the Occaney gap [354 623], becoming lower (25 m above OD) towards Staveley. The terrace

appears to be later than the Nidd terraces, described above, and to have been deposited either when the Devensian ice front was stationary at a position north of the Flaxby moraine (Figure 27), or when thick glacial deposits blocked the valley here, so that the easterly drainage along Robert Beck was blocked and the water overflowed through the Copgrove channel. To the north-east, this terrace interdigitates with lacustrine deposits, whereas to the south-west it passes abruptly into laminated clay (Abraham, 1981), which overlies the Devensian till present in the buried valley at Farnham.

Another fluvioglacial terrace deposit, possibly also related to a still stand of the ice sheet, occurs at Deighton Grange [410 516] (Figure 27). This terrace has an elevation of around 25 m above OD, its northern limit is sharp, and it extends southwards for about 4 km into the adjacent district. It joins up with terrace deposits emanating from the Kirk Deighton glacial drainage channel and probably formed when the ice front was stationary just north of Deighton Grange; in this position the ice would also have blocked the easterly drainage, causing it to utilise the Kirk Deighton channel. Just south of the Harrogate district, at Loshpot Lane [4098 5067], a mineral assessment borehole showed the deposit to comprise clayey, pebbly sand up to 4.3 m thick (Giles, 1981).

### The Ripon deposits

A third group of deposits is present in the vicinity of Ripon, where there are extensive fluvioglacial terraces along the rivers Skell and Ure (Figure 29). Adjacent to the River Skell, at Borrage Green Gravel Pit [3065 7055], the terrace has an elevation of around 38 m above OD. The deposit is over 7 m thick and comprises yellow-brown, clayey, sandy gravel which has cross-bedding and well-developed channel structures. The gravel ranges from pebble to boulder size and has the composition of about 40 per cent Carboniferous limestone, 30 per cent Carboniferous sandstone and 30 per cent Permian limestone (Cooper, 1983; Morigi and James, 1984). This deposit reduces in height eastwards, through Ripon, to around 25 m in the vicinity of Littlethorpe. At the same time, the grain size decreases from sand and gravel southwards to sand around Givendale [340 690], then into sandy silt and ultimately into laminated clay of glacial lake origin (elevation about 24 m above OD) south of Littlethorpe and on Roecliffe Moor (see below).

## Glacial Lake Deposits

As the Devensian ice sheet stagnated and the ice-front retreated northwards, the meltwater formed widespread glacial lakes, mainly impounded behind the moraines running across the Vale of York. Fluvioglacial terrace deposits formed by the meltwaters pass distally into sand, silt and clay laid down in the glacial lakes. Sand and gravel, as marginal strand lines and beach deposits, are particularly widespread where the glacial lake deposits abut against sandy till or earlier glacial sand and gravel. The glacial lake deposits have a distinctive flat morphology, commonly curved gently upwards around the edges on to the lake-margin topography; till and glacial sand and gravel features sporadically protrude upwards through the lake deposits. Scattered hollows probably indicate sites where blocks of ice were buried beneath the lake sediments and subsequently melted.

### Silt and clay

Silt and clay are the most abundant glacial lake sediments; they form the surface deposit over appoximately one-quarter of the Harrogate district. They range from a feather edge to 16 m in thickness and, where seen in section, consist of dark grey and brown clay, in layers generally 1 to 4 mm thick, interlaminated with light brown silt layers, generally less than 1 mm thick. This lithology is typical of varved deposits, representing annual cycles of sedimentation, the silt deposited from low-density suspension currents during the spring meltwater influx, the clay settling-out of fine suspended material in winter (Sturm and Matter, 1978). The lamination is generally subhorizontal, but may also form drapes over local irregularities, such as sporadic ripple-marked beds of fine-grained sand. However, in Littlethorpe clay pit [3264 6811], the lamination (Plate 11) is tilted at angles up to 10°, and the lake deposits are cratered by numerous subsidence hollows caused by foundering of the bedrock following dissolution of Permian gypsum (see chapter eight).

The largest lake deposit, located behind the York moraine in the south-east of the district, forms a flat area of heavy clay soil which extends eastwards a further fifteen kilometres into the York (Sheet 63) district. Its surface ranges in elevation from 13 to 16 m above OD, although marginal deposits towards Easingwold attain an elevation up to 27 m above OD. The area between Roecliffe Moor [365 655], west of Boroughbridge, and Littlethorpe, near Ripon, also marks the site of another glacial lake, the surface of which also ranges in elevation from 15 to 27 m above OD; this lake appears to have been connected to the main eastern lake via a channel at Boroughbridge (Figure 29).

Other isolated deposits of laminated clay, in the central part of the district, formed in localised lakes ponded up by the glacial deposits. The largest of these are at Farnham [343 605] (surface level up to 40 m above OD), Arkendale [380 600] (up to 48 m above OD) and Goldsborough Moor [395 550] (up to 38 m above OD).

Small areas of laminated clay, around the confluence of the River Nidd and Oak Beck, apparently mark the site of a short-lived lake ponded up by the glacial diversion of the River Nidd, and subsequently drained via Knaresborough gorge.

### Details

Near Littlethorpe, laminated clay deposits are dug from a clay pit [3264 6811] to manufacture garden pottery at Littlethorpe pottery [3248 6813]. In the clay pit, the section exposes up to 3m of brown clay with laminae of brown and pale brown silt. In the workings, the laminae are mainly horizontal (Plate 11), but in disused clay pits nearby [3276 6790], undulating laminae which turned down into enclosed depressions were reported.

These depressions were probably caused by dissolution of Permian gypsum; active subsidence of this type has been recorded in the vicinity (see chapter eight).

In the north-east of the district, 800 m west of Far Shires, glacial lake clays are exposed in Alne pit, where the clay has been excavated for brick making; in 1977, the working face [5203 6635] showed the following section:

| | *Thickness* m |
|---|---|
| Sandy clay topsoil | 0.40 |
| Yellow sandy clay | 0.50 |
| Brown to greenish grey, well-laminated, stone-free clay with ill-defined bands of grey-green clay | 4.60 |
| Dark brown, laminated, stone-free clay with thin beds of light brown silt | 0.50 |

Information from the pit manager indicated that lacustrine clay extends a further 2.5 m below the base of the section and rests on till. He also reported that cobbles and boulders are very occasionally found in the laminated clay; these may be interpreted as dropstones (ice-rafted erratics floated out into the proglacial lake and released when the ice melted). Also in the north-east of the district, boreholes at the former Easingwold Flax Works [5263 6725] and at Forest Farm [5053 6739] showed the glacial lake clays to be 7.2 m and possibly 13.1 m thick respectively. In a nearby flooded clay pit at Burnside Farm [5487 6562] stone-free clays were reported to a depth of 10 m below ground level.

In the south-east of the district, a water pipeline excavation exposed an almost continuous section of the glacial lake deposits from near Tockwith [4766 5236] to near Red House [5255 5723]. Along most of its length, the excavation was entirely in the laminated clay to a depth of 3 to 4 m, but at its southern end the underlying till was also proved. Lenticular sand deposits were also exposed; these are discussed below. A typical section illustrating the sequence was seen on Moor Monkton Moor [5085 5553 to 5050 5516]. Here, the area is flat, with a heavy clay soil to a depth of 0.3 m on sandy clay and laminated clay to a depth of around 1.3 m. Below this, impersistent lenticular beds of orange-brown, fine- to medium-grained sand up to 0.5 m thick and 3 to 4 m across rest on homogeneous, dark brown, laminated clay to a depth of up to 4 m. On Marston Moor [4821 5288], the pipeline excavation exposed grey-brown clay with laminae of silt and sand plus sporadic sand lenticles to 2.5 m, resting on grey-brown clay with sporadic pebbles (till) to a depth of 3.5 m.

### Sand and gravel

Sand and gravel of glacial lake origin occurs in two main situations; as marginal beach or strand-line deposits, and as deposits distal to fluvioglacial terraces. However, the two types occur in close association and generally they cannot be separated.

Clayey sands along the eastern edge of Roecliffe Moor [381 640] are of possible strand-line origin and reach 1.1 m in thickness (Abraham, 1981); similar deposits occur mainly along the east side of Arkendale Moor [387 601]. Such lake-margin deposits were probably formed by current or wave action driven by the dominant westerly winds, hence their occurrence along the eastern sides of the former lakes. Around Myton on Swale [450 670], Linton on Ouse [496 608] and from Tholthorpe [475

**Plate 11** Laminated clay (Glacial Lake Deposits) at Littlethorpe Pottery [3264 6811], near Ripon. This varved clay, which contains laminae of fine silt, is hand-dug for the manufacture of garden pottery (L 3036).

675] to Easingwold [535 685], broad spreads of fine-grained clayey sand are interbedded with the lacustrine silts and clays; generally 1 to 2 m thick, they reach a maximum of 6 m (Stanczyszyn, 1982). They may represent reworking by wave action of sand from the edge of the tills, and of glacial sand and gravel; however, their widespread extent suggests that they may have been derived from glacial meltwater entering the lake in the vicinity.

In the south of the district, sands and gravels of this type are restricted to the area where the River Nidd flowed into the lake. In the west, around Walshford [419 529] and Cowthorpe [426 528], they form spreads of pebbly sand up to 1.5 m thick; these pass eastwards into lenticular, distal deposits of fine-grained sand, mainly concentrated around Marston Moor [485 534] and Tockwith [475 526]. On Marston Moor, the sands form low ridges elevated up to 1 m above the surrounding clay flat. In section, many of these ridges have a lenticular form and also cut down into the underlying clay. They may represent distributary channels with levees, formed after the glacial lake drained, but before the true fluvial drainage routes were established.

## Details

Sand and gravel of lacustrine origin is present where the River Nidd enters the low ground of the Vale of York lake near Cowthorpe. A section adjacent to the river [4423 5294] showed the following section:

| | Thickness m |
|---|---|
| Sandy soil, brown and very pebbly | 0.15 |
| Sandy gravel with abundant pebbles and cobbles of hard sandstone in a matrix of pale brown coarse-grained sand | 0.40 |
| Sand, pale brown, coarse-grained | 0.10 |
| Sand, orange, medium- and coarse-grained, with faint subhorizontal bedding and a poorly exposed uneven bottom contact | 0.90 |
| Clay, blue-grey and grey-brown, homogeneous in the top 0.4 m, but becoming laminated in the bottom part | 1.40 |
| No exposure (landslip) | 1.50 |
| Clay, blue-grey and grey-brown, laminated and poorly exposed | 0.60 |
| No exposure | 1.10 |
| Sandstone, orange-red, coarse-grained, thick-bedded, with cross-bedding (Sherwood Sandstone Group) | 1.45 |

Nearby, a pipeline trench [4406 5269] exposed a section with a slightly different sequence and penetrating the underlying till:

| | Thickness<br>m |
|---|---|
| Sand and topsoil | 0.30 |
| Clay, brown, laminated, wedging out slightly<br>  northwards | 1.00 |
| Sand, orange-brown | 3.00 |
| Clay, brown, laminated | 1.50 |
| Sandy clay (till), grey-brown, with pebbles and<br>  cobbles | 2.50 |

The Eccup to Red House water pipeline trench [4637 5122 to 5255 5723] exposed sand and gravel of glacial lake origin in several places. At the contact of the glacial lake deposits and the underlying till near Tockwith [4782 5247], strand-line deposits occur, the section seen through these being:

| | Thickness<br>m |
|---|---|
| Clayey sand with sporadic pebbles | 0.30 |
| Clay, pale brownish white to orange-brown, sandy | 0.40 |
| Clay, dark grey, laminated, with silt laminae | 1.00–1.10 |
| Clay, dark brown, and silt, brownish white, both<br>  laminated; also laminae and very thin beds of<br>  fine-grained, red-brown sand, including some<br>  beds with slight ripple-drift cross-lamination | 0.90–1.00 |
| Sand, red-brown, fine- to medium-grained, with<br>  an irregular contact on the underlying till | 0.15–0.50 |
| Sandy clay (till), dark brown, with abundant<br>  small pebbles, cobbles and boulders of<br>  sandstone, and several angular fragments of<br>  Magnesian Limestone | 0.30–0.50 |

A ditch section on Wilstrop Moor [4875 5377] showed the sand hereabouts to have the form of a channel fill. The sand, which is slightly clayey and fine to coarse grained, stands up as ridges up to 1 m higher than the surrounding flat of laminated clay deposits. In the ditch section, the sand was seen to be up to 1.5 m thick and to have an erosional base cutting down into the laminated clay. This suggests a fluvial origin for the sand, probably contemporaneous with the draining of the glacial lake and the initiation of fluvial drainage.

## River Terrace Deposits

As the Devensian glaciation drew to a close, the ice melted from the Vale of York and surrounding areas, the glacial lakes drained away and the present drainage patterns became established, although these were largely controlled by the pre-existing glacial features. The rivers Swale and Ure flowed across, and incised into the low-lying glacial lake deposits. The River Nidd upstream was confined by its gorge, but downstream cut through the fluvioglacial terrace deposits and till to spill onto the glacial lake deposits and join the Ouse drainage. The Ouse also followed the established glacial drainage pattern and breached the York moraine at York, near the line of the esker between the York and Eskrick moraines.

As the River Nidd cut down through the glacial deposits to its new base level, the glacial material was reworked and deposited as terraces, which occur at a level below that of the fluvioglacial terrace deposits, but above that of the present floodplain alluvium. Terraces occur adjacent to the River Nidd alluvium from the south-eastern end of Knaresborough gorge to as far east as Tockwith. Terraces are absent where the river has traversed the glacial lake deposits, presumably because the gradient of the river was already low and less downcutting occurred. This observation may also explain the lack of terraces adjacent to the rivers Ure and Ouse in the Harrogate district, as these rivers run mainly across glacial lake deposits.

### Details

The terraces adjacent to the River Nidd consist mainly of sand and gravel with a similar composition to that of the glacial deposits from which they were derived. They generally occur as flattish benches flanking the river alluvium a few metres above the present floodplain. At the exit of Knaresborough gorge, the river terrace [3695 5647] has an elevation of around 35 m above OD and comprises slightly clayey pebbly sand with sporadic boulders. North of Little Ribston, the terrace has an elevation of 26.7 m and comprises 2.8 m of clayey silt and sandy clay on at least 3.5 m of clayey gravel, as proved in a mineral assessment borehole [3822 5441] (Dundas, 1981). Farther downstream, the terrace, also proved in a borehole [3990 5310] by Dundas (1981), comprises 1 m of silty clay resting on at least 0.5 m of mainly clayey sandy gravel. In the vicinity of Tockwith and Cattal, a terrace proved by a mineral assessment borehole [4448 5367] has an elevation of 19.8 m above OD, and comprises 1.8 m of mainly clayey, pebbly sand resting on glacial lake deposits (Giles, 1981). East of Cattal, another borehole [4599 5396], on a terrace with an elevation of 15.2 m above OD, proved sandy clay to 1.6 m resting on very clayey sandy gravel to 5.0 m (Giles, 1981). The elevation of this terrace is only about 2.0 m below that of the surrounding glacial lake deposits and only about 1.0 m above the present river alluvium; no terraces occur adjacent to the River Nidd to the east of here.

The River Swale has one terrace only [427 692], near Burton Grange, with an elevation of around 14 to 15 m above OD. It largely comprises brown sandy silt, proved to a depth of 2.5 m in a mineral assessment borehole [4256 6944] near Burton Grange, and silty clay proved to 3.4 m in another assessment borehole [4215 6824] near Humburton (Stanczyszyn, 1982).

### Blown sand

Blown sand in the Harrogate district is almost exclusively associated with fine sand of glacial lake origin, from which much of it was derived; till and glacial sand and gravel probably also provided a minor source. Active derivation of blown sand from glacial lake deposits was observed around Coneythorpe in 1976 [390 590] where fine-grained glacial lake sands (in freshly harrowed fields on Arkendale Moor) were picked up by a westerly gale and spread on the west-facing slope below Coneythorpe.

The widespread blown sand plastered on the foot of the Jurassic escarpment of Easingwold [550 685] was probably formed in a similar manner, but was built up gradually after the glacial lake drained. Along the Stillington road [539 689], south-east of Easingwold, blown sands form a series of hummocks and ridges generally elongated in an north-north-westerly direction; the low ground between these ridges is occupied by head (see below). Farther south, around Hall Bank Farm [553 606], grey to brown and orange-brown, blown silty sand occurs as a series of low, elongate, north-west-trending, dune-like ridges, generally less than 1.5 m thick. Similar blown sands at East Moor [608 640], in the York area, overlie peat which yields radiocarbon ages of 10 700 to 9950 years BP (Matthews, 1970).

### Details

At Easingwold, a temporary section [5282 6959] in Long Street was reported to have penetrated 1.5 m of fine-grained sand. At a building site [532 698] adjacent to Kellbalk Lane, trial pits and boreholes proved between 1.1 and 1.6 m of brown and grey, fine- to medium-grained, moderately sorted blown sand resting on glacial sand. Exposures in Westmoor Drain [5498 6861; 5491 6841] showed between 1.1 and 2.3 m of yellow, orange and brownish white, well-sorted very fine-grained sand with signs of even lamination.

### Head

Deposits classified as Head comprise thin, locally derived, stony and sandy clays which have moved down slopes by the action of solifluxion (periglacial freeze and thaw action). The deposits accumulated to form even-surfaced sheets of superficial material on slopes and in valley bottoms. Being ill-sorted deposits, they are commonly difficult to distinguish from till in areas where the latter has contributed to their source material. Head is not normally mapped in these situations. Mappable head is widespread on the lower slopes of the Jurassic escarpment around Easingwold [540 695], overlying blown sand; it is also mappable as stony clay across a flat valley floor at Poppleton [550 544].

### Details

A small patch of head forms a flat area of poorly drained ground [529 703] some 500 m north-east of Easingwold Church. Farther east, a narrow strip of head occupies the floor of the valley of Haverwits Beck [5405 7020], but to the east and south [540 695] of Cottage Farm this expands to cover a broad area of flattish ground which extends south-eastwards towards Hollins Grove. Soils hereabouts are markedly clayey and augering has proved up to 0.9 m of mainly sandy clay. North-west of Cottage Farm, narrow irregular tongues of head, lying between slight ridges of blown sand, extend downslope from the Jurassic scarp to the north. West of Long Bridge [5365 6911], tongues of head merge into a broad flat area where up to 1 m of clay and sandy clay rests on glacial lake sands. East [548 688] of Hollins Grove, a broad tongue of head, which gives rise to sandy clay soils with sparse pebbles, extends south-westwards from a narrow valley in the Jurassic scarp to die out at the northern edge of the glacial lake sands on Ox Moor. An exposure [5498 6861] in Westmoor Drain shows 0.8 m of dark brown, slightly sandy clay resting on blown sand.

### Calcareous tufa

In the valley of Holbeck, to the north of Copgrove, [345 643] sponge-like calcareous tufa containing sporadic gastropod shells is associated with peat. The tufa occurs both as thin beds and irregular masses, which give rise to small hillocks up to 2 m high covered by a thin veneer of peat. These deposits are related to springs emanating from the Magnesian Limestone and gypsiferous marls of the Permian bedrock. Similar tufa deposits associated with springs have been recorded from the Burton Salmon area of West Yorkshire (Norris et al., 1971). At Ripon racecourse, calcareous springs well up beneath the gravels, causing them to become partially cemented by calcareous tufa (Abraham, 1981, p.8, table 2, borehole NW 12).

Other deposits of calcareous tufa, with shell marl, are commonly interbedded in peat deposits. Near Dallamires Lane [318 703], Ripon, beds of shell marl up to 0.15 m thick occur with peat filling a subsidence area caused by gypsum dissolution.

In the north-east of the district, shell marl, though not exposed at the surface, is inferred to underlie blown sand to the north-west of Easingwold, from the evidence of material brought up from the re-excavation [5174 7029] of a field drain. The section had been filled in by the time of the survey, but fragments of soft, pale grey clay with numerous small molluscs were collected from the spoil. Mr D K Graham identified the following fauna: *Columella edentula?*, *Lymnaea truncatula*, *Oxyloma pfeifferi*, *Succinea oblonga*, *Valvata cristata*, *Vertigo pygmaea?*, succineid fragments, vertiginid fragments, unidentifiable gastropod fragments and *Pisidium* sp..

Calcareous tufa is also present at the Dropping Well [3479 5650], Knaresborough where it forms a tufa screen associated with a prolific spring; the formation of this is discussed in chapter eight.

### Alluvium

With the exception of the River Nidd along the Knaresborough gorge, all the rivers in the Harrogate district have alluvial floodplains. These low-lying, flat areas are up to 1.5 km wide and are prone to flooding, except where artificial levees have been constructed. The deposits present beneath the flat are normally in two layers. The lower layer, generally less than 3.5 m thick, is a lag deposit of sand and gravel left by meandering point bars; this sand and gravel has a similar provenance to much of the glacial deposits from which it was originally derived. Overbank or flood deposits, forming the upper layer are up to 6 m thick. They consist of brown, thinly bedded to laminated, silty clays containing sporadic beds and lenses of fine-grained sand. Adjacent to the rivers, these silt and clay deposits are prone to landslipping. Sporadic lenses and patches of peat are present in the alluvium, especially on the sites of abandoned meanders.

Numerous ill-drained hollows also contain alluvial clay and silt, commonly associated with peat. The majority are in areas mantled by till and some are typical kettle holes. Other hollows, largely confined to areas underlain by the Permian Middle Marl, Upper Magnesian Limestone or Upper Marl are subsidence features resulting from the removal, by dissolution, of gypsum from the marl formations.

### Details

In the vicinity of Ripon racecourse, alluvial sand and gravel is worked adjacent to the River Ure. The sequence here [3296

6902] comprises an upper deposit of clay about 1 m thick, underlain by sandy clay, then yellow-brown clayey sand. This rests on the gravel which is usually about 3.6 m thick, but both the base and the top of the deposit rise and fall in level when traced laterally. In places, the gravel deposit is cut by channel deposits of clay, sand and peat, with shell marl. West of Ripon racecourse, a mineral assessment borehole [3272 6961], elevation 18.7 m above OD, proved silty clay to 2.1 m, over slightly sandy gravel to 5.2 m, resting on glacial lake deposits of laminated clay (Abraham, 1981). Downstream near Mulwith Farm, another borehole [3599 6641], elevation 14.9 m above OD, proved sandy clay to 2.5 m, on very clayey sand to 4.0 m, resting on fluvioglacial clay.

East of Boroughbridge, the alluvial deposits of the rivers Ure, Swale and Ouse have been penetrated by numerous mineral assessment boreholes (Stanczysyn, 1982); these have mainly penetrated silt and clay to between 4.0 and 6.0 m depth, with only subordinate lag deposits of gravel and sand. Adjacent to the River Nidd, at the exit from Knaresborough gorge, the river alluvium, proved in a borehole [3675 5661] with a surface elevation of 32.5 m above OD, comprised silty and sandy clay to 4.0 m on sandy gravel to 6.1 m (Dundas, 1981). Downstream near Little Ribston, the alluvium proved in another mineral assessment borehole [3960 5366] comprised silty sand to 0.7 m, on clayey gravel to 3.5 m, resting on till (Dundas, 1981). A similar sequence was also proved for the Nidd alluvium in the vicinity of Kirk Hammerton, where another mineral assessment borehole [4845 5632], elevation 10.0 m above OD, penetrated sandy clay to 3.5 m, on clayey, pebbly sand to 7.1 m, on red Sherwood Sandstone (Giles, 1981).

## Peat

Peat is present in numerous low-lying and ill-drained flat areas throughout the district. It also occurs in kettle holes, abandoned meanders and gypsum subsidence hollows, often interbedded with alluvium. The largest deposit is north-east of Staveley [374 638], where peat to a depth of up to 1.6 m was proved in two mineral assessment boreholes [3722 6416; 3700 6366] (Abraham, 1981). Large areas of peat also occur at Derrings [465 685] and Thorpe Underwoods [458 593]. In the Holbeck valley, peat is associated with springs and occurs in association with calcareous tufa. At Dallamires Lane, Ripon [318 703], a flat-surfaced deposit of peat and laminated clay reaches a thickness of 30 or 40 m in a large uneven-bottomed subsidence crater (see chapter eight), and a similar peat-filled area occurs near Coneythorpe [386 587].

## Landslip

Landslips commonly develop on steep slopes, notably where streams and rivers have eroded into the foot of unstable deposits. Only a few areas of landslip are large enough to show on the published map. The main areas occur in Knaresborough gorge and adjacent to the River Nidd near Goldsborough (Plate 12). Further details are given in chapter eight (p.82).

## Made ground

Made ground in the district is very variable both in thickness and lithology. Around Aldborough [405 665], an extensive area of made ground is largely of Roman origin, being the remains of the Roman town of Svrivm. Near Knaresborough, made ground associated with the Farnham gravel pits [363 580] comprises a mixture of domestic waste and quarry spoil. Around Harrogate, the numerous small areas of made ground also include domestic waste and spoil from pipeline excavations. The deposit at the disused Tockwith airfield [460 523] is mainly material scraped from the runway areas during construction. Made ground can provide difficult ground conditions for development (see chapter eight).

**Plate 12**   Landslip developed in the Middle Marl near to the River Nidd at Goldsborough Mill [3699 5608] (L 1802).

# EIGHT

# Economic geology

## RESOURCES

### Mineralisation

Marshall (1826) and Sedgwick (1829) recorded that malachite, both disseminated in the country-rock and as 'potato-sized masses', was worked from the Lower Magnesian Limestone (Cadeby Formation) near Farnham. Around 1765, a 'considerable quantity' of ore was mined from shafts and galleries (14 to 15 m deep), probably on Folly Hill [349 608], but the operation was abandoned due to flooding and to disputes that prevented the installation of a drainage system. The flora of the limestone soil around Folly Hill suffers from excess copper, and soil samples analysed by H V Warren (personal communication, 1980) yielded up to 1600ppm of the metal. Copper is present in the soil as far south as Gibbet Hill [356 598] and as far north as Walkingham Hill [343 619], the mineralisation being localised in the area where the Ripley–Staveley Fault Belt intersects the Burton Leonard–Tollerton Monocline. These structures probably relate to basement faults, and a deep-rooted source of the mineralisation is probable (Harwood and Smith, 1986).

The Lower Magnesian Limestone at Wormald Green also hosts mineralisation, which includes calcitised displacive anhydrite nodules with sphalerite, marcasite, barite and galena (Harwood, 1980; Harwood and Smith, 1986). The mineralisation is of penecontemporaneous origin, formed by bacterial reduction of evaporites during diagenesis (Harwood and Colman, 1983).

### Carboniferous rocks

Siltstones and mudstones of Carboniferous age were formely worked for the manufacture of bricks in the vicinity of Harrogate. The two largest quarries were at Stonefall [331 548] and Grange [2865 5770]; numerous small pits were also excavated. All these workings have now been filled in and brickmaking has ceased in the area.

The thicker sandstones were in the past extensively quarried for local building stone, both for houses and for the dry-stone walls that are a feature of the countryside along the western side of the district. The Harrogate Roadstone (p.11) was formerly worked from many quarries all along its narrow outcrop, partly for road-metal and partly for walling.

### Magnesian Limestone

The Lower Magnesian Limestone (Cadeby Formation) has been extensively worked in the Harrogate district; details of the abandoned and working quarries are given by Cooper (1987). The rock is mainly dolomite, but sporadic beds of calcitic dolomite are also present. Rock Cot-

tage Quarry, Wormald Green [300 650], was, at the time of the survey, the only working quarry in the district. Here, approximately the middle 24 m of the Lower Magnesian Limestone were worked for agricultural limestone and low grade roadstone aggregate. Large resources of the Lower Magnesian Limestone, only thinly covered by drift, are available in the district.

The Upper Magnesian Limestone (Brotherton Formation) is more calcitic, but is thinner, so it offers less potential as an aggregate source. All the old quarries in this formation are small and disused; details of them are given in Cooper (1987). The largest quarry, exposing the best section in the formation, is near Burton Leonard [3319 6345] (Plate 7).

### Gypsum

Sequences of gypsum are present both in the Permian Middle and Upper Marls, where they are up to about 30 m and 10 m thick respectively. The gypsum belt is from 2 to 5 km wide and extends from the outcrop down dip to the gypsum–anhydrite transition (see chapter three for details). The gypsum is secondary, formed by the near-surface hydration of anhydrite; however, the interaction between the groundwater and gypsum eventually causes it to dissolve, with resulting ground subsidence. The widespread occurrence of subsidence and the existence of a considerable groundwater flow into the deep buried valley, extending from near Ripon [326 706] southwards to the east of Burton Leonard [346 643], suggest that gypsum has been extensively attacked by dissolution and is unlikely ever to be workable in this area. South-eastwards, from Burton Leonard to the southern boundaries of the district, subsidence hollows are more scattered and little active subsidence has been recorded; it is probable that groundwater flow is less in this area and the gypsum beds may be of economic importance.

### Sand and gravel

The widespread sand and gravel deposits in the Harrogate district were intensely investigated by the Geological Survey Industrial Minerals Assessment Unit, which prepared detailed resource reports for much of the area (Abraham, 1981; Dundas, 1981; Giles, 1981; Stanczyszyn, 1982). Around Farnham [349 600] and Knaresborough [363 582], sand and gravel is worked from beneath the Devensian till in the buried valley system. Considerable resources remain, but the deposit becomes finer grained and more clayey both to the north-west and to the southeast of the present workings, and the thickness of overburden also increases.

Glacial sand and gravel was formerly quarried from numerous small pits, mainly for local use. The only major excavations were made in the deposits of the Helperby–Aldwark esker at the now disused pits of Ten Mile Hill [4658 6714], Flawith [4747 6615] and Aldwark [4609 6473]. Recently, glacial sand and gravel has been exploited from a new pit [406 600] in the Hunsingore esker near Allerton Park. Fluvioglacial terrace deposits were formerly worked at Staveley [366 635], Haughs Farm [366 563], Ripley [290 603] and Ripon [306 705]. Sand and gravel at the base of the alluvium, together with older gravel deposits immediately beneath, is extracted at Ripon Race Course [330 695].

## Clay

In the past, laminated clay deposits of glacial lake origin were widely exploited for local brickmaking; disused clay pits of medium size exist at Goldsborough Moor [3907 5641] and near Tockwith [4564 5372; 472 532]. Numerous small clay pits occur scattered throughout the district, as it was a common practice for clamps of bricks to be made by itinerant brickmakers near the sites of proposed buildings. Until fairly recently, bricks and tiles were manufactured at Roecliffe [385 660], where there are extensive disused clay pits. The only working clay pits in the district are near Littlethorpe Pottery [3248 6813] (Plate 11), where handmade decorative gardenware is produced, and at Alne Brickworks [5193 6630], which at one time produced agricultural drainage materials and currently makes handmade bricks.

## Hydrocarbons

The Harrogate district has been the subject of oil and gas exploration intermittently since 1945, although few details relating to hydrocarbon prospectivity have been published. Two exploration wells have been drilled, at Ellenthorpe [4233 6703] in 1945 by D'Arcy Exploration Co. Ltd (BP) (Appendix 1; Falcon and Kent, 1960; Allsop, 1985) and at Tholthorpe [4683 6690] in 1965 by Home Oil of Canada Ltd. Both were dry, as were four others drilled nearby in adjacent districts. Between the years 1973–1987 an open grid of 300 to 400 line-km of seismic reflection data have been acquired in the district by a series of exploration companies. Regional reviews of hydrocarbon prospectivity in northern England, including the Harrogate district, have been published by Kirby et al. (1987) and Fraser et al. (1990).

The main potential hydrocarbon source rocks are contained in the Carboniferous sequence. Shales with suitable (moderate to good) organic carbon content (0.1 per cent) are probably present throughout the Carboniferous succession and coal seams are relatively numerous in Westphalian rocks above the middle part of the Lower Coal Measures. The organic matter is dominantly terrestrially derived, and is believed to be mainly gas-prone, commonly with only poor hydrocarbon generating potential. However, some oil-prone shales are likely to be present in the early Pendleian Upper Bowland Shales (equivalent to the beds above the Harrogate Roadstone), although the thickness of these may well prove to be restricted. The concealed Westphalian A/B coals, in the east of the district, are high volatile bituminous coals indicative of the upper part of the oil window. Up to the present time, they can have sourced only relatively minor amounts of gas. Making due allowance for the differing burial histories of Carboniferous rocks across the district, the deltaic and marine Pendleian and Arnsbergian shales are probably generally in, or just below the lower part of the oil-generating zone. Such rocks could yield gas, wet gas and perhaps some light oils. More deeply buried, basinal, Dinantian shales are probably totally within the gas-generating zone.

Any hydrocarbons generated during late Carboniferous burial probably escaped during the Hercynian earth movements and the subsequent period of extensive subaerial sub-Permian erosion. Except in the west, the maximum burial of early Namurian rocks, and related hydrocarbon generation, probably occurred during late Cretaceous times. Potential reservoirs and trapping structures were probably present at this time, both in the Hercynian folds and the fault-bounded closures in the Carboniferous rocks, and also in younger rocks. However, if, as believed, the district is essentially gas-prone, it is uncertain whether such structures could have been properly sealed. Fraser et al. (1990) have suggested that the Carboniferous rocks of North Yorkshire may provide only poor gas seals, and it also seems likely that the marginal Permian rocks, with only relatively thin evaporite sequences and lacking thick halite beds, would also provide poor gas seals. Tertiary faulting, uplift and erosion have further reduced the chances of any successful preservation of sealed hydrocarbon-bearing accumulations.

The only post-Carboniferous rocks with source potential are the immature bituminous shales of the Whitby Mudstone Formation, which have only a restricted occurrence near the north-east margin of the district. Here, as elsewhere regionally, these are believed to have yielded no significant hydrocarbon occurrences.

In summary, the district appears to have only limited potential for hydrocarbons, mainly gas. However, so little is known of the stratigraphy, lithology and structure of the concealed Carboniferous rocks of the central and eastern parts of the district, that this conclusion may prove to be premature.

# HYDROGEOLOGY

## Water supply and springs

The district is drained by the River Ure and its tributary, the River Swale, which pass downstream into the River Ouse with its tributary, the River Nidd. Due largely to the extensive cover of clay-rich drift, there is only a small groundwater contribution to surface flow, and the rivers tend to suffer flash flow, with recorded minimum flows generally less than 10 per cent of the mean. The mean annual rainfall ranges from over 700 mm in the north-west to about 650 mm in the south-east. The mean annual evapotranspiration is of the order of 400 mm, and there is also a considerable loss by run-off to streams, so that the mean annual infiltration into the aquifers does not appear to exceed 150 to 190 mm.

There are no major reservoirs in the district and most of the water supplies are obtained from a river intake on the Ouse at Moor Monkton [527 576], which extracts up to 68 000 cubic metres per day ($m^3/d$). Additional supplies are also taken from wells and boreholes, as detailed below, and from sources outside the district.

Hydrogeologically, the district may be divided into three unequal areas, broadly comprising the Carboniferous strata in the west, the Lower and Upper Magnesian limestones in the middle and the large, mainly drift-covered outcrop of the Sherwood Sandstone Group in the centre and east.

In both the Carboniferous and Permian areas, groundwater storage and flow is mainly within fissures in the more massive sandstones and magnesian limestones respectively. Boreholes intersecting such fissures provide groundwater sources, but amounts pumped are usually for domestic or small agricultural purposes. Exceptionally, a borehole at Burton Leonard [3205 6378] pumps up to 125 $m^3/d$ from the Lower Magnesian Limestone for public supply. The groundwater quality in these areas is generally good. The total hardness usually ranges from 300 to 500 mg/l; the chloride ion concentration is normally less than 50 mg/l, and sulphate (as $SO_4$) rarely exceeds 150 mg/l. The Lower Magnesian Limestone aquifer is rarely used where confined beneath the overlying Permian formations and the Sherwood Sandstone, because the quality there tends to deteriorate; sulphate may exceed 1000 mg/l. Where the gypsiferous sequences of the Middle and Upper Marls are penetrated by boreholes, the sulphate concentration in the ground water may also be considerable. Near Ripon, two boreholes [3199 7058; 3199 7055] through the Middle Marl gypsum sequence yield water with sulphate contents of 1200 and 1500 mg/l. The volume of water abstracted from these boreholes was around 1200 $m^3/d$ in 1979; the volume of sulphate being removed with the groundwater is considerable and may have implications for gypsum dissolution and potential subsidence (see below and Cooper, 1988).

In the Sherwood Sandstone, groundwater flow may be dominantly through fissures, but there is a significant intergranular storage. There are numerous small sources taking groundwater for agricultural and domestic use. Two major public supply sources, at Lower Dunsforth [435 643] and Marton-cum-Grafton [422 632], together pump up to 4500 $m^3/d$; a smaller source at Kirby Hill [387 683] yields up to 250 $m^3/d$. Boreholes at Tholthorpe [469 670] and Upper Dunsforth [449 632] abstract up to 3300 $m^3/d$ for river flow augmentation. The total hardness of groundwater from the Sherwood Sandstone is generally between 100 and 300 mg/l, tending to increase beneath the overlying Mercia Mudstone in the north-east of the district; the chloride ion concentration is usually less than 100 mg/l, while sulphate rarely exceeds 250 mg/l. Groundwater is generally abstracted from excavated shafts or, more commonly at the present day, from boreholes. In both the Lower and Upper Magnesian Limestones and the Sherwood Sandstone, a bore-

hole will normally stand unsupported and requires only a few metres of casing below rockhead to prevent the direct inflow of potential contaminants.

Springs are relatively uncommon in the Harrogate district and are usually of small size, issuing from small patches of permeable drift such as glacial sands and gravels. Along the course of Holbeck from Mongah's Well [3470 6379] northwards numerous fairly prolific springs [3446 6411; 3452 6436; 3451 6462] emanate from the Permian sequence and well up through the thick drift deposits; the springs are associated with areas of peat and raised areas of calcareous tufa deposited from the hard spring water. In the vicinity of Ripon Racecourse, the gravel deposits in the buried valley contain sulphate-rich groundwater and are cemented with calcareous tufa (Abraham, 1981); this suggests prolonged spring activity in the area, but the spring water is generally lost into the alluvial sands and gravels and does not appear at the surface.

### Harrogate mineral springs

Within a three kilometre radius of Harrogate, 94 springs, most of which yield mineral waters, issue from Carboniferous strata (Jennings, 1974); their distribution is closely related to the geological structure (Figure 30). The springs have previously been classified under four headings: saline sulphur, alkaline sulphur, saline iron and pure chalybeate (Fox-Strangways, 1908). The majority of the springs are located along the axis of the asymmetrical Harrogate Anticline and are most abundant where this approaches the Harrogate Fault (Figure 30). The waters issue principally from a near-vertical, interbedded sandstone–shale sequence of late Brigantian and early Pendleian age, chiefly the strata between the Harlow Hill Sandstone and the Harrogate Roadstone.

Most of the waters appear not to be in hydraulic continuity with one another. Variation in their chemistry bears a definite relation to the structure: those issuing near the axis of the anticline have higher total mineralisation (maximum 16 500 mg/l). The principal ions are $Na^+$ and $Cl^-$; $HCO_3^-$ and dissolved sulphide ($HS^-$) concentrations may be high (Table 1, 1950 analysis). Some of the waters also contain barium (up to 68 mg/l). The composition of the most saline spring (Old Sulphur Well) has remained constant for at least 100 years, as has that of many springs; others, mainly those more remote from the axis of the anticline, show considerable seasonal fluctuations (Smithells, 1919).

The relationship between the major cations in 18 of the Harrogate Springs is shown in Figure 31, after Edmunds et al. (1969), from which it can be seen that:

1 There is a relatively constant Mg/Ca ratio.
2 Waters high in dissolved solids plot in a restricted area.
3 Waters with a low total dissolved solid content also show the same cationic composition as waters high in dissolved solids.

The more remote and least saline sources may be derived by mixing of surface waters with saline water having

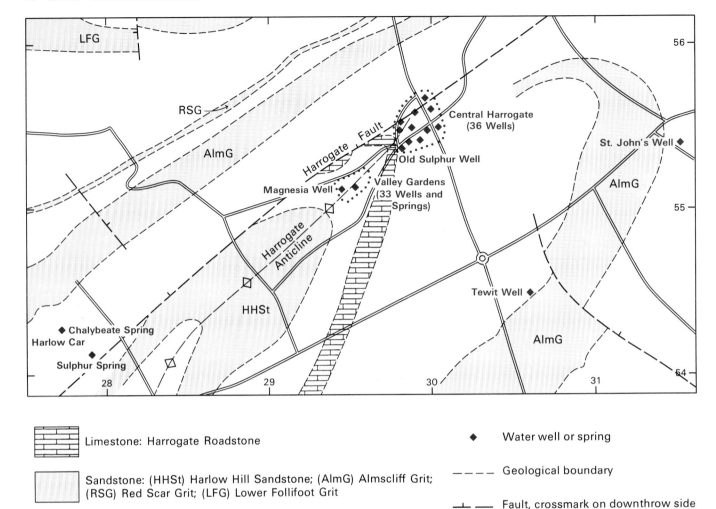

**Figure 30**  Location of the principal Harrogate mineral springs and wells, related to bedrock geology.

a Mg/Ca ratio slightly less than 1. The oxygen and hydrogen stable isotope ratios ($^{18}O$ and H) indicate that the (saline) water is of meteoric, not marine origin and has an isotopic composition similar to that of Bath thermal water, which also has a meteoric origin (Figure 32) (Andrews et al., 1982) the source of the salinity may therefore be evaporite NaCl dissolved by meteoric circulation. The temperature of the most saline Harrogate water is only 14°C, which implies either derivation from shallow depth or a very slow ascent from a deeper source. Calculation of the maximum subsurface temperature using silica concentrations as a geothermometer also gives an equilibrium temperature with quartz of 14°C, implying a shallow depth of circulation.

Thus, the modern geochemical evidence is consistent with Hudson's view (1938b) that the saline waters probably derived from a limestone-shale sequence near the base of the Millstone Grit sequence. The high sulphur concentration could result from in-situ reduction of sulphate, metabolised by bacteria under strongly reducing conditions; this would also account for the high $HCO_3^-$ derived from oxidation of organic carbon, and for the low content of iron in the waters, due to precipitation of iron sulphide according to the reaction:

$$15CH_2O + 2Fe_2O_3 + 8SO_4^{2-} + H_2CO_3 = 4FeS_2 + 16HCO_3^- + 8H_2O$$

The presence of barium and sulphate in the Old Sulphur Well water is anomalous and probably indicates some mixing of sulphate-free and sulphate-bearing waters to give rise to the saline waters. The more remote and less saline spring waters are derived from mixing of surface water with the mineral spring waters.

**Table 1** Chemistry of Harrogate mineral spring waters: composition in mg/l; see Figure 30 for the location of the springs.

| | Old Sulphur Well (inside) | | Old Sulphur Well (outside) | Magnesia | St John's |
|---|---|---|---|---|---|
| Date | 13/11/85 | c.1950 | 13/11/85 | 14/11/85 | 27/11/85 |
| pH (laboratory) | 8.1 | — | 8.2 | 8.1 | 7.9 |
| Na | 5090 | 5180 | 5030 | 945 | 15.4 |
| K | 58 | 69 | 57 | 14.4 | 1.6 |
| Ca | 410 | 430 | 180 | 38 | 10.2 |
| Mg | 180 | 191 | 180 | 38 | 10.2 |
| $HCO_3$ | — | 1222 | 339 | 360 | 112 |
| $SO_4$ | 129 | 2 | 152 | 9.1 | 15.3 |
| Cl | 8900 | 8928 | 8900 | 1630 | 17.5 |
| Sr | 26 | 16 | 23 | 5.6 | 0.083 |
| Ba | 52 | 68 | 2.5 | 30 | 0.001 |
| B | 2.3 | — | 2.4 | 0.62 | 0.035 |
| Si | 4.3 | — | 5.5 | 6.6 | 13.7 |
| Li | 3.1 | 1.2 | 3.0 | 0.61 | 0.010 |
| Fe | <0.1 | tr | 0.18 | 0.06 | 7.5 |
| Mn | 0.20 | 0.30 | 0.33 | 0.46 | 0.41 |
| F | 2.4 | — | 2.3 | 8.2 | 0.17 |
| I | 0.810 | 0.1 | 0.360 | 0.064 | 0.006 |
| Br | | 28 | | | |
| $HS^-$ | | 89.6 | | | |
| Total determined mineralisation | 14858 | 16225 | 14878 | 3087 | 204 |
| $\delta^{18}O$ | -54 | — | -52 | -51 | -54 |
| $\delta^2H$ | -7.6 | — | -7.2 | -7.2 | -7.9 |

## The Dropping Well, Knaresborough

Adjacent to the River Nidd at Knaresborough [3479 5650], the Dropping Well spring issues from the Permian Middle Marl. The spring is heavily laden with dissolved sulphate and carbonate (Table 2 and Edmunds et al., 1969) and has deposited an extensive screen of calcareous tufa which has built out from the valley side. This screen is about 5 m high and approximately 10 m wide; the water runs in channels across the top and cascades down the face depositing tufa as it flows. The spring water has a petrifying nature and is used to petrify objects for the delight of tourists. The adjacent Mother Shipton's Cave is also formed beneath an older tufa

screen and indicates that the major outlet of the spring has moved with time. In the catchment area to the south and west of the spring, there are several enclosed hollows with ponds; these are probably subsidence craters formed by the dissolution of the gypsum in the Middle Marl (see below). The water analysis for the Dropping Well (Table 2) also shows the presence of strontium (Sr), which in the original analysis is given as strontium sulphate; this is probably produced by the dissolution of small amounts of celestine, a mineral which occurs in the Permian sequence, and was recorded from the banks

**Table 2** Chemistry of mineral water from the Dropping Well, Knaresborough; composition in mg/l.

| | |
|---|---|
| Na | 15.5 |
| K | 3.5 |
| Ca | 633.2 |
| Mg | 49.1 |
| $HCO_3$ | 223.6 |
| $SO_4$ | 1572.1 |
| Cl | 27.1 |
| Sr | 4.6 |
| Si | 4.9 |
| Fe | 1.1 |
| Mn | 0.6 |
| Total determined mineralisation | 2535.5 |

Date 1896 (recalculated from Fox-Strangways, 1908; partly after Edmunds et al., 1969).

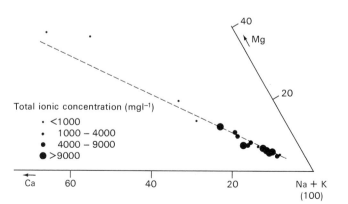

**Figure 31** Major cation compositions of 18 Harrogate waters.

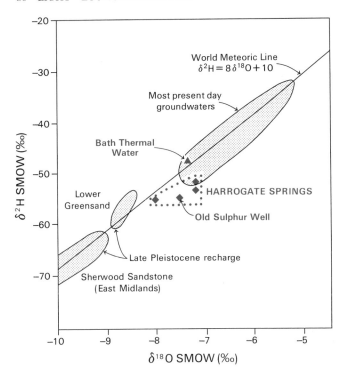

**Figure 32** Plot of hydrogen and oxygen stable isotope data for the Harrogate mineral waters relative to SMOW (Standard Mean Ocean Water) and their relationship to other groundwaters.

of the River Nidd near Bilton (Murray, 1817; Platnauer, 1888; Fox-Strangways, 1908).

## PLANNING CONSIDERATIONS

Within the district there are several potential sources of geological hazard which are outlined below. Any development should be preceded by the seeking of specialist engineering geological and geotechnical advice, and by an adequate site investigation. In particular, investigations within the area shown by Figure 33 should determine the presence or absence of subsidence features caused by gypsum dissolution; the sulphate contents of both the soil and the groundwater should also be determined to acertain if sulphate-resistant concrete is required. The design of the foundations should allow for these problems. Other potential problems in the district include landslips, weak compressible soils and made ground; these also require close investigation. The problem area can then be avoided, or special foundations can be designed to cope with it.

### Subsidence caused by the dissolution of gypsum

Gypsum dissolves in flowing water about one hundred times more rapidly than limestone: a block of Middle Marl gypsum about 3 m across was dissolved almost completely in about 14 months by the River Ure at Ripon Parks (James et al., 1981). Where there is a considerable

groundwater flow, the Permian gypsum beds have dissolved, resulting in caverns that periodically collapse, so forming subsidence hollows and foundered strata. The distribution of the subsidence belt (Figure 33) is limited to the west by the outcrop of the base of the gypsum beds and to the east by the down-dip transition of gypsum to the much less soluble anhydrite (see chapter three). The region between Nosterfield [280 805], to the north of the district, Ripon [313 713] and Bishop Monkton [332 663] has suffered about forty instances of subsidence in the past 150 years, some of them sudden and catastrophic (Cooper, 1986; 1989).

The area most prone to subsidence is centred on Ripon and extends south to Bishop Monkton. In this area, unusual hydrological conditions are caused by the presence of a deep buried valley intersecting the Permian and Triassic formations. The Magnesian Limestones act as a catchment area for water, which moves down dip and into the adjacent gypsum beds. It escapes from this sequence into the buried valley, dissolving the gypsum beds as it passes through them. Evidence of this water movement also comes from the calcareous tufa around springs near Burton Leonard [345 644], from tufa-cemented gravels in the buried valley (Abraham, 1981; Morigi and James, 1984) and from sulphate-saturated groundwater in boreholes near Ripon Race Course (Cooper, 1986). A similar association of gypsum dissolution, sulphate-rich water and calcareous tufa deposition occurs at the Dropping Well [3479 5650], described above.

The phreatic flow of groundwater through the rock dissolves the gypsum, mainly along the joints. This has produced a cave system in the gypsum, with caverns at the intersections of the joints (Cooper, 1986). As the dissolution of the gypsum continues at a rapid rate, the caverns enlarge, amalgamate, become unstable and ultimately collapse. When failure of the cavity roof occurs, a breccia pipe may work its way up to the surface causing subsidence features to develop (Cooper, 1988). These range from a slight sagging of the ground surface (subsidence sinkholes of Culshaw and Waltham, 1987) to small crown holes with large voids beneath, through to complete catastrophic collapse (collapse sinkholes of Culshaw and Waltham, 1987). The larger subsidences may be up to 20 m deep and 30 m across. Some of the large failures have been felt as very local earthquakes caused by the roof plug of rock dropping into the cavity.

The physical nature of the major subsidence hollows (collapse sinkholes of Culshaw and Waltham, 1987) depends on the surface geology. Cylindrical shafts generally form where rock is present at the surface, but in superfical deposits failure of the sides causes conical depressions to develop. These subsidence features occur in a linear or grid-like pattern directly related to the joint-controlled cave system beneath (Figure 34 and Cooper, 1986; 1989).

Once initiated, the cave systems appear to maintain some water flow even after a collapse has occurred. Consequently, areas adjacent to, or in line with known subsidence hollows are more at risk from future subsidence.

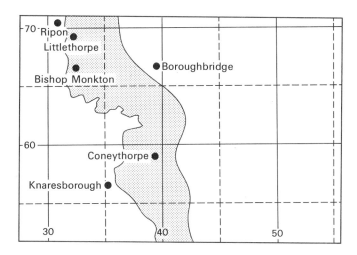

**Figure 33** Approximate limits of area subject to foundering due to the groundwater dissolution of gypsum and anhydrite in the Middle and Upper Marls.

One of the most spectacular subsidences recorded in the district was at Littlethorpe [3245 6891] (Figure 34) where 'On 16th October 1796 a violent earthquake hit Ripon; at Littlethorpe three roods of ground sunk nineteen fathoms and a large ash tree entirely disappeared into the hole' (Harrison, 1892). Another major subsidence occurred around 1830 at Bishop Monkton [?3375 6566] (Figure 34) (Tute, 1868; 1870), and in 1980 a subsidence at Dallamires Lane, Ripon [3187 7043] was recorded by Cooper (1986). The large peat- and clay-filled depression at the latter site (Figure 34) has been formed by repeated subsidence and represents an amalgamation of numberous subsidence hollows. The surface of the area is flat, but the bottom of the the depression is a highly irregular series of intersecting conical depressions, each one representing a subsidence hollow. The infill deposits range in thickness from a few metres to as much as 30 or 40 m.

In the vicinity of Littlethorpe Pottery [3248 6813], eastwards to Ox Close House [3311 6810] and south-west towards Moor End [3193 6783], deposits of silt and clay formed in glacial lakes are affected by gypsum dissolution subsidence. Numerous enclosed hollows up to 50 m in diameter occur in the otherwise flat deposits, and sections in the pottery clay pits in the vicinity commonly show the subhorizontal varve laminations bending downwards into the hollows. The subsidence hollows around here occur in lines, which have approximately north-west and north-east-trending orientations, similar to those in the Ripon area (Cooper, 1986). Within a radius of approximately 2 km around Bishop Monkton [330 664], the area is similarly dotted with enclosed hollows (Figure 34), but here they affect glacial till, the surface of which naturally tends to have enclosed hollows (kettle holes) of glacial origin. However, many of the hollows around Bishop Monkton also occur in linear belts. This, and the fact that active subsidence has occurred in the area, suggests that many of them may be attributed to subsidence caused by gypsum dissolution. Farther south, the hydro-

logical conditions are less favourable for the dissolution of the gypsum, but numerous enclosed hollows may also be attributable to this agency. Uncertainty exists for the origin of those in the till-covered areas, but those affecting flat lake deposits, such as the low area of peat near Coneythorpe [386 587], are certainly subsidence features. Other evidence of foundering caused by gypsum dissolution is given by the variable dips observed in the Upper Magnesian Limestone at Burton Leonard [3319 6345] and Ribston Hall [3900 5400] (see details in chapter three).

**Sulphate-rich groundwater**

The subsurface dissolution of gypsum (see above) causes the groundwater present in the bedrock to be rich in dissolved sulphate. This water commonly issues from springs and is also present in the drift deposits, as at Ripon Racecourse (Cooper, 1986). Sulphate-rich groundwater has the potential to harm buried concrete, and sulphate-resistant concrete may need to be used following the guidelines of Anon (1981). Where a proposed structure includes the use of buried concrete, the sulphate contents of the soil and groundwater should be measured as part of the site investigation. The area most likely to be affected is similar to the area prone to subsidence caused by gypsum dissolution (Figure 33).

**Landslips**

Extensive areas of landslip occur along the steep banks of the River Nidd at Bilton Hall [334 576], Gallow Hill [346 566] and Goldsborough [371 561; 376 554] (Plate 12). At Bilton Hall, landslips affect Carboniferous mudstones and the complete Permian sequence, which has been disrupted by movements on the Middle and Upper Marls. Landslip in other areas south-eastwards along the Nidd valley is less extensive and confined to the Middle Marl; the overlying Upper Magnesian Limestone is commonly cambered slightly towards the river in these cases. Laminated clays (glacial lake deposits) are also prone to landslip development, especially adjacent to rivers or in excavations. The published 1:10 000 and 1:50 000 scale geological maps do not differentiate landslip types. Most of the slips are rotational in their upper parts, with flows developed on the lower slopes; however, detailed classification was beyond the scope of the survey and each area of landslip must be fully investigated before any development is carried out on or adjacent to it. Caution must be exercised with respect to any excavations or modifications of surface drainage, as these may cause reactivation of dormant landslips.

**Weak compressible soils (laminated clay, alluvium and peat)**

Glacial deposits of laminated silt and clay may present problems of poor ground conditions to foundation engineers. These deposits are generally evenly bedded and have a high water content. Their shear strength is low and they are highly compressible; thus problems of ex-

**Figure 34**   Distribution of enclosed hollows (coloured brown) in the area between Ripon and Bishop Monkton. The great majority are attributed to subsidence resulting from dissolution of gypsum in the Middle and Upper Marls; a few may be glacial kettle holes, but where till is at the surface these are difficult to differentiate from subsidence hollows.

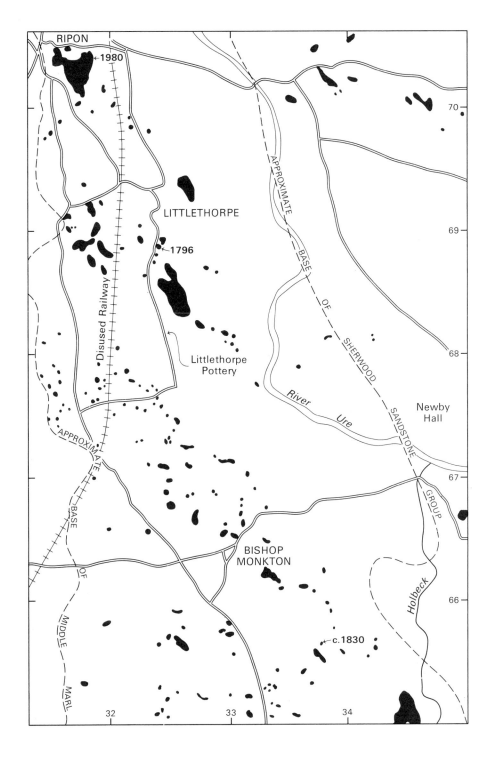

cessive settlement and bearing capacity failure may occur. The common occurrence of interbedded, fine-grained, water-bearing sands may lead to instability even on gentle slopes; these sands may also run into excavations, which therefore require close shuttering. Heave, caused by the unloading of the laminated clays, or due to artesian water below thin clay at the bottom of excavations, may also be a problem.

The silt and clay overbank deposits in the alluvium of river and stream valleys may also behave in a similar manner to the glacial lake clays, and interbedded peat layers may pose additional problems. These are very variable deposits and each site must be investigated carefully.

Peat has a high water content, a very low bearing-capacity and is highly compressible. It can, therefore, cause severe problems of excessive and differential settlement in engineering works. This is especially so in subsidence hollows where it may locally be up to 30 or 40 m thick, and in areas where it forms extensive spreads up to a few metres thick. Acidic groundwater in the peat may also be detrimental to concrete, metal reinforcements and metal services.

## Made ground

The distribution of the made ground is described in chapter seven. Areas of made ground should be treated with caution. They commonly include domestic refuse and poorly compacted fill leading to difficult foundation conditions. Furthermore, the migration from the landfill sites of methane-rich gases, caused by the breakdown of organic matter, can lead to potentially dangerous situations where porous bedrock or drift deposits occur in the vicinity.

# REFERENCES

Most of the references listed below are held in the Library of the British Geological Survey at Keyworth, Nottingham. Copies of the references can be purchased subject to the current copyright legislation.

ABRAHAM, D A. 1981. The sand and gravel resources of the country west of Boroughbridge, North Yorkshire. Description of 1:25 000 resource Sheet SE 36. *Mineral Assessment Report, Institute of Geological Sciences*, No. 78.

ACRILL, R. LTD. 1935. Old Harrogate. Photographs published in the Harrogate Herald 1934–1935.

ALLSOP, J M. 1985. Geophysical indications of the sub-Permian geology beneath the Ripon area, Northern England. *Proceedings of the Geologists' Association*, Vol. 96, 161–169.

ANDERTON, R, BRIDGES, P H, LEEDER, M R, and SELLWOOD, B W. 1979. A dynamic stratigraphy of the British Isles. 301pp. (London: Allen & Unwin.)

ANDREWS, J N, BURGESS, W G, EDMUNDS, W M, KAY, R L F, and LEE, D J. 1982. The thermal springs of Bath. *Nature, London*, Vol. 298, 339–343.

ANON. 1981. Concrete in sulphate-bearing soils and groundwaters. *Building Research Establishment Digest*, No. 250.

ARTHURTON, R S. 1983. The Skipton Rock Fault — an Hercynian wrench fault associated with the Skipton Anticline, northwest England. *Geological Journal*, Vol. 18, 105–114.

— 1984. The Ribblesdale Fold Belt, northwest England — a Dinantian–early Namurian dextral shear zone. 131–146 *in* Variscan tectonics of the North Atlantic region. SANDERSON, D, and HUTTON, D (editors). *Special Publication of the Geological Society of London*, No. 14.

— JOHNSON, E W, and MUNDY, D J C. 1988. Geology of the country around Settle. *Memoir of the British Geological Survey*, Sheet 60 (England and Wales). 147pp.

BANERJEE, I, and McDONALD, B C. 1975. Nature of esker sedimentation. 132–154 *in* Glaciofluvial and glaciolacustrine sedimentation. JOPLIN, A V, and McDONALD, B C (editors). *Special Publication of the Society of Economic Paleontologists and Mineralogists*, No. 23.

BENFIELD, A C. 1983. The geology of the country around Dalton, North Yorkshire, with particular reference to sand and gravel deposits: description of 1:25 000 Sheet SE 47. *Open File Report, Institute of Geological Sciences*. 28pp.

— and WARRINGTON, G. 1988. New records of the Westbury Formation (Penarth Group, Rhaetian) in North Yorkshire, England. *Proceedings of the Yorkshire Geological Society*, Vol. 47, 29–32.

BOTT, M H P. 1978. Deep structure. 25–40 in *The geology of the Lake District*. MOSELEY, F (editor). 384pp. (Leeds: Yorkshire Geological Society.)

BOWEN, D Q. 1978. Quaternary geology. 221pp. (Oxford: Pergamon Press Ltd.)

— ROSE, J, McCABE, A M, and SUTHERLAND, D G. 1986. Correlation of Quaternary glaciations in England, Ireland, Scotland and Wales. 299–340 *in* Quaternary glaciations in the northern hemisphere. SIBRAVA, V, BOWEN, D Q, and RICHMOND, G M (editors). *Quaternary Science Reviews*, Vol. 5.

BRISTOW, C S. 1988. Controls on the sedimentation of the Rough Rock Group (Namurian) from the Pennine Basin of northern England. 114–131 in *Sedimentation in a synorogenic basin complex: the Upper Carboniferous of northwest Europe*. BESLEY, B M, and KELLING, G (editors). 276pp. (Blackie: Glasgow and London.)

BRITISH GEOLOGICAL SURVEY. 1983. York: England and Wales Sheet 63. Solid and Drift. 1:50 000. (Southampton: Ordnance Survey for the British Geological Survey).

—. Masham: England and Wales Sheet 51. Solid. 1:50 000. (Southampton: Ordnance Survey for the British Geological Survey.)

BURGESS, I C, and COOPER, A H. 1980a. The Permian and Carboniferous rocks of the Knaresborough district; description of field meeting 16th September 1978. *Proceedings of the Yorkshire Geological Society*, Vol. 43, 36–38.

— 1980b. The Farnham (IGS) borehole near Knaresborough, North Yorkshire. *Report of the Institute of Geological Sciences*, No. 80/1, 12–17.

CHISHOLM, J I. 1981. Growth faulting in the Almscliff Grit (Namurian $E_1$) near Harrogate, Yorkshire. *Transactions of the Leeds Geological Association*, No. 9, 61–70.

CHUBB, L J, and HUDSON, R G S. 1925. The nature of the junction between the Lower Carboniferous and the Millstone Grit of north-west Yorkshire. *Proceedings of the Yorkshire Geological Society*, Vol. 20, 257–291.

CLARK, D N. 1980. The diagenesis of Zechstein carbonate sediments. 167–203 *in* The Zechstein Basin with emphasis on carbonate sequences. FUCHTBAUER, H, and PERYT, T M (editors). *Contributions to Sedimentology*, No. 9.

COLLINSON, J D. 1988. Controls on Namurian sedimentation in the Central Province basins of northern England. 83–101 in *Sedimentation in a synorogenic basin complex: the Upper Carboniferous of northwest Europe*. BESLEY, B M, and KELLING, G (editors). 276pp. (Blackie: Glasgow and London.)

COLTER, V S, and REED, G E. 1980. Zechstein 2 Fordon Evaporites of the Atwick OB. 1 borehole, surrounding areas of NE England the adjacent southern North Sea. 115–129 *in* The Zechstein Basin with emphasis on carbonate sequences. FUCHTBAUER, H, and PERYT, T M (editors). *Contributions to Sedimentology*, No. 9.

COOPER, A H. 1983. The geology of the country north and east and Ripon, North Yorkshire, with particular reference to the sand and gravel deposits; description of 1:25 000 Sheet SE 37. *Open File Report, Institute of Geological Sciences*. 33pp.

— 1986. Foundered strata and subsidence resulting from the dissolution of Permian gypsum in the Ripon and Bedale area, North Yorkshire. 127–139 *in* The English Zechstein and related topics. HARWOOD, G M, and SMITH, D B (editors). *Special Publication of the Geological Society of London*, No. 22.

— 1987. The Permian rocks of the Harrogate district; geological description and local details of 1:50 000 Sheet 62 and component 1:10 000 Sheets SE 25 NE, SE 26 NE/SE, SE 35

NW/SW/NE/SE, SE 36 NW/SW/SE and SE 45 SW. *Open File Report, British Geological Survey.* 49pp.

— 1988. Subsidence resulting from the dissolution of Permian gypsum in the Ripon area; its relevance to mining and water abstraction. 387–390 in Engineering geology of underground movements. BELL, F G, CULSHAW, M G, CRIPPS, J C, and LOVELL, M A (editors). *Special Publication of the Geological Society of London*, No. 5.

— 1989. Airborne multispectral scanning of subsidence caused by Permian gypsum dissolution at Ripon, North Yorkshire. *Quarternary Journal of Engineering Geology, London,* Vol. 22, 219–229.

COPE, J C W, GETTY, T A, HOWARTH, M K, MORTON, N, and TORRENS, H S. 1980. A correlation of Jurassic rocks in the British Isles. Part I: Introduction and Lower Jurassic. *Special Report of the Geological Society of London*, No. 14.

CULSHAW, M G, and WALTHAM, A C. 1987. Natural and artificial cavities as ground engineering hazards. *Quarterly Journal of Engineering Geology, London*, Vol. 20, 139–150.

DEANS,T. 1950. The Kupferschiefer and associated mineralisation in the Permian of Silesia, Germany and England. *Report of the XVII International Geological Congress* (London). Pt.7, 340–352.

DUNDAS, D L. 1981. The sand and gravel resources of the country east of Harrogate, North Yorkshire. Description of 1:25 000 Sheet SE 35. *Mineral Assessment Report, Institute of Geological Sciences*, No. 76.

DUNHAM, K C. 1974. Granite beneath the Pennines in North Yorkshire. *Proceedings of the Yorkshire Geological Society*, Vol. 40, 191–194.

— and WILSON, A A. 1985. Geology of the North Pennine Orefield. Vol. 2 Stainmore to Craven. *Economic Memoir of the British Geological Survey*, Sheets 40, 41 and 50 and parts of 31, 32, 51, 60, 61 (England and Wales).

EBDON, C C, FRASER, A J, HIGGINS, A C, MITCHENER, B C, and STRANK, A R E. 1990. The Dinantian stratigraphy of the East Midlands: a seismostratigraphic approach. *Journal of the Geological Society of London*, Vol. 147, 519–536.

EDMUNDS, W M, TAYLOR, B J, and DOWNING, R A. 1969. Mineral and thermal waters in the United Kingdom. *Proceedings of the XXXII International Geological Congress*, Vol. 18, 139–158.

EDWARDS, W. 1936. Pleistocene dreikanter in the Vale of York. *Summary of Progress of the Geological Survey for 1934*, Part 2, 8–19.

— 1938. The geology of the country around Harrogate; part IV, The glacial geology. *Proceedings of the Geologists' Association*, Vol. 49, 333–343.

— 1951. The concealed coalfield of Yorkshire and Nottinghamshire (3rd edition). *Memoir of the Geological Survey of Great Britain.* 285pp.

— MITCHELL, G H, and WHITFIELD, T H. 1950. Geology of the district north and east and Leeds. *Memoir of the Geological Survey of Great Britain*, Sheet 70 (England and Wales). 93pp.

— WRAY, D A, and MITCHELL, G H. 1940. Geology of the country around Wakefield. *Memoir of the Geological Survey of Great Britain*, Sheet 78 (England and Wales). 215pp.

FALCON, N F, and KENT, P E. 1960. Geological results of petroleum exploration in Britain 1945–57. *Memoir of the Geological Society of London*, No. 2. 56pp.

FOX-STRANGWAYS, C. 1874. The geology of the country north and east of Harrogate. *Memoir of the Geological Survey,*

Explanation of Quater-sheet 93 NW (England and Wales). 21pp.

— 1885. The Harrogate Wells, or the mineral waters of Harrogate geologically considered. *Proceedings of the Yorkshire Geological and Polytechnic Society*, Vol. 8 (for 1884), 319–337.

— 1908. The geology of the country north and east of Harrogate (2nd edition). *Memoir of the Geological Survey*, Sheet 62 (England and Wales). 100pp.

— CAMERON, A C G, and BARROW, G. 1886. The geology of the country around Northallerton and Thirsk. *Memoir of the Geological Survey*, Explanation of Quarter-sheets 96 NW and 96 SW (England and Wales). 75pp.

— and BARROW, G. 1915. The geology of the country between Whitby and Scarborough. *Memoir of the Geological Survey*, Explanation of Quarter-sheet 95 NW (England and Wales).

FRASER, A J, NASH, D F, STEELE, R P, and EBDON, C C. 1990. A regional assessment of the intra–Carboniferous play of northern England. 417–439 in Classic petroleum provinces. BROOKS, J (editor). *Special Publication of the Geological Society of London*, No. 50. 568pp.

FUZESY, L M. 1980. Origin of nodular limestones, calcium sulphates and dolomites in the Lower Magnesian Limestone in the neighbourhood of Selby, Yorkshire, England. 35–44 in The Zechstein Basin with emphasis on carbonate sequences. FUCHTBAUER, H, and PERYT, T M (editors). *Contributions to Sedimentology*, No. 9.

GAUNT, G D. 1970. An occurrence of Pleistocene ventifacts at Aldborough, near Boroughbridge, West Yorkshire. *Journal of Earth Sciences*, Vol. 8, 159–161.

— 1976. The Devensian maximum ice limit in the Vale of York. *Proceedings of the Yorkshire Geological Society*, Vol. 40, 631–637.

— 1981. Quaternary history of the southern part of the Vale of York. 82–97 in *The Quaternary in Britain*. NEALE, J, and FLENLEY, J (editors). (Oxford: Pergamon Press.)

— BARTLEY, R D, and HARLAND, R. 1974. Two interglacial deposits proved in boreholes in the southern part of the Vale of York and their bearing on contemporaneous sea levels. *Bulletin of the Geological Survey of Great Britain*, No. 48, 1–23.

— JARVIS, R A, and MATTHEWS, B. 1971. The Late Weichselian sequence in the Vale of York. *Proceedings of the Yorkshire Geological Society*, Vol. 38, 281–284.

GAWTHORPE, R L. 1987. Tectono-sedimentary evolution of the Bowland Basin, N England, during the Dinantian. *Journal of the Geological Society of London*, Vol. 144, 59–71.

GEORGE, T N. 1932. Brachiopoda from the Cayton Gill beds. *Transactions of the Leeds Geological Association*, Vol. 5, 37–48.

— JOHNSON, G A L, MITCHELL, M, PRENTICE, J E, RAMSBOTTOM, W H C, SEVASTOPULO, G D, and WILSON, R B. 1976. A correlation of the Dinantian rocks in the British Isles. *Special Report of the Geological Society of London*, No. 7.

GILES, J R A. 1981. The sand and gravel resources of the country around Kirk Hammerton, North Yorkshire: description of 1:25 000 resource sheet SE 45. *Mineral Assessment Report, Institute of Geological Science*, No. 84.

GILLIGAN, A. 1920. The petrography of the Millstone Grit of Yorkshire. *Quarterly Journal of the Geological Society of London*, Vol. 75, 251–294, pls. 15–18.

GLENNIE, K W. 1982. Early Permain (Rotliegendes) Palaeowinds of the North Sea. *Sedimentary Geology*, Vol. 34, 245–265.

— 1983. Lower Permian Rotliegend desert sedimentation in the North Sea area. 521–541 in Eolian sediments and processes. BROOKFIELD, M E, and AHLBRANDT, T S (editors). *Developments in Sedimentology*, No. 38. 660pp.

GODWIN, C G. 1973. The geology of the Ure valley water main trenches 1969–71. *Proceedings of the Yorkshire Geological Society*, Vol. 39, 537–546.

GUION, P D, and FIELDING, C R. 1988. Westphalian A and B sedimentation in the Pennine Basin, UK. 151–177 in *Sedimentation in a synorogenic basin complex: the Upper Carboniferous of northwest Europe*. BESLEY, B M, and KELLING, G (editors). 276pp. (Blackie: Glasgow and London.)

HARLAND, W B. 1971. Tectonic transpression. *Geological Magazine*, Vol. 108, 27–42.

— ARMSTRONG, R L, COX, A V, CRAIG, L E, SMITH, A G, and SMITH, D G. 1989. *A geological timescale 1989.* 260pp. (Cambridge University: Cambridge Press.)

HARMER, F E. 1928. The distribution of erratics and drift. With a contoured map. *Proceedings of the Yorkshire Geological Society*, Vol. 21, 79–150.

HARRISON, W (editor). 1982. *Ripon Millenary. A record of the festival also a history of the city arranged under its wakemen and mayors from the year 1400.* 629pp. (Ripon: W. Harrison.)

HARWOOD, G M. 1980. Calcitized anhydrite and associated sulphides in the English Zechstein first cycle carbonate (EZ1 Ca). 61–72 in The Zechstein Basin with emphasis on carbonate sequences. FUCHTBAUER, H, and PERYT, T M (editors). *Contributions to Sedimentology*, No. 9.

— and COLMAN, M L. 1983. Isotopic evidence for U.K. Upper Permian mineralization by bacterial reduction of evaporites. *Nature, London*, Vol. 301, 597–599.

— and SMITH, F W. 1986. Mineralization in Upper Permian carbonates at outcrop in eastern England. 103–111 in The English Zechstein and related topics. HARWOOD, G M, and SMITH, D B (editors). *Special Publication of the Geological Society of London*, No. 22.

HEDBERG, H D. 1976. *International stratigraphic guide.* 200pp. (New York: J Wiley.)

HEMINGWAY, J E. 1974. Jurassic. 161–223 in *The geology and mineral resources of Yorkshire*. RAYNER, D H, and HEMINGWAY, J E (editors). 405pp. (Leeds: Yorkshire Geological Society.)

— and RIDDLER, G P. 1982. Basin inversion in North Yorkshire. *Transactions of the Institute of Mining and Metallurgy, (Section B: Applied Earth Sciences)*, Vol. 91, 175–186.

HOLDSWORTH, B K, and COLLINSON, J D. 1988. Millstone Grit cyclicity revisited. 132–152 in *Sedimentation in a synorogenic basin complex: the Upper Carboniferous of northwest Europe*. BESLEY, B M, and KELLING, G (editors). 276pp. (Blackie: Glasgow and London.)

HUDLESTON, W H. 1883. Excursion to the West Riding of Yorkshire. *Proceedings of the Geologists' Association*, Vol. 7, 420–438.

HUDSON, R G S. 1930. The Lower Carboniferous of the Harrogate Anticline. *Transactions of the Leeds Geological Association*, Part XX, 33–40.

— 1934. The Millstone Grit succession south of Harrogate. *Transactions of the Leeds Geological Association*, Vol. 5, 118–124.

— 1938a. The geology of the country around Harrogate; Part II; the Carboniferous rocks. *Proceedings of the Geologists' Association*, Vol. 49, 306–330.

— 1938b. The geology of the country around Harrogate; Part VI; the Harrogate mineral waters. *Proceedings of the Geologists' Association*, Vol. 49, 349–352.

— 1939. The Millstone Grit succession of the Simonseat Anticline, Yorkshire. *Proceedings of the Yorkshire Geological Society*, Vol. 23, 319–349.

— 1941. The Mirk Fell Beds (Namurian E$_2$) of Tan Hill, Yorkshire. *Proceedings of the Yorkshire Geological Society*, Vol. 24, 259–289.

— 1944. The faunal succession in the *Ct. nitidus* Zone in the mid-Pennines. *Proceedings of the Leeds Philosophical and Literary Society, (Science Section)*, Vol. 4, 233–242.

— EDWARDS, W, TONKS, L H, and VERSEY, H C. 1938. Report of summer field meeting 1938: The Harrogate district. *Proceedings of the Geologists' Association*, Vol. 49, 353–372.

— and MITCHELL, G H. 1937. The Carboniferous geology of the Skipton Anticline. *Summary of Progress of the Geological Survey of Great Britain*, (for 1935), Part II, 1–45.

INGRAM, R L. 1954. Terminology for the thickness of stratification and parting units in sedimentary rocks. *Bulletin of the Geological Society of America*, Vol. 65, 125–145.

INSTITUTE OF GEOLOGICAL SCIENCES. 1971. *Annual report for 1970.* (London: Institute of Geological Sciences.)

— 1973. Pickering: England and Wales Sheet 53. Solid and Drift 1:50 000. (Southampton: Ordnance Survey for the Institute of Geological Sciences.)

— 1976. *Annual report for 1975.* (London: Institute of Geological Sciences.)

— 1976. IGS boreholes 1975. *Report of the Institute of Geological Sciences*, No. 76/10. 46pp.

— 1977. 1:625 000 Series. Quaternary map of the United Kingdom, South.

— 1983. IGS boreholes 1982. *Report of the Institute of Geological Sciences*, No. 83/11. 10pp.

IVIMEY-COOK, H C, and POWELL, J H. 1991. Late Triassic and early Jurassic biostratigraphy of the Felixkirk Borehole, North Yorkshire. *Proceedings of the Yorkshire Geological Society*, Vol. 48, 367–374.

JAMES, A N, COOPER, A H, and HOLLIDAY, D W. 1981. Solution of the gypsum cliff (Permian Middle Marl) by the River Ure at Ripon Parks, North Yorkshire. *Proceedings of the Yorkshire Geological Society*, Vol. 43, 433–450.

JENNINGS, B. 1974. *A history of the wells and springs of Harrogate.* 27pp. (Harrogate: Harrogate Corporation Department of Conference and Resort Services.)

JOHNSON, G A L, HODGE, B L, and FAIRBAIRN, R A. 1962. The base of the Namurian and of the Millstone Grit in north-eastern England. *Proceedings of the Yorkshire Geological Society*, Vol. 33, 341–362.

KALDI, J G. 1980a. Aspects of the sedimentology of the Lower Magnesian Limestone (Permian) of Eastern England. Unpublished PhD thesis, University of Cambridge. 239pp.

— 1980b. The origin of nodular structures in the Lower Magnesian Limestone (Permian) of Yorkshire, England. 45–60 in The Zechstein Basin with emphasis on carbonate sequences. FUCHTBAUER, H, and PERYT, T M (editors). *Contributions to Sedimentology*, No. 9.

— and GIDMAN, J. 1982. Early diagenetic dolomite cements: examples from the Permian Lower Magnesian Limestone of

England and the Pleistocene carbonates of the Bahamas. *Journal of Sedimentology and Petrology*, Vol. 52, 1073–1085.

— and SWALLOW, P W. 1987. Tectonism and sedimentation in the Flamborough Head region of north-east England. *Proceedings of the Yorkshire Geological Society*, Vol. 46, 301–309.

KENDALL, P F, and WROOT, H E. 1924. *The geology of Yorkshire*. 995pp. (Vienna: privately printed.)

KENT, P. 1966. The structure of the concealed Carboniferous rocks of north-eastern England. *Proceedings of the Yorkshire Geological Society*, Vol. 35, 323–352.

— 1974. Structural history. 13–30 in *The geology and mineral resources of Yorkshire*. RAYNER, D H, and HEMINGWAY, J E (editors). 405pp. (Leeds: Yorkshire Geological Society.)

— 1980. *British regional geology: Eastern England from the Tees to the Wash* (2nd edition). 155pp. (London: HMSO.)

KIRBY, G A, SMITH, K, SMITH, N P J, and SWALLOW, P. 1987. Oil and gas generation in eastern England. 171–180 in *Petroleum geology of north-west Europe*. BROOKS, J, and GLENNIE, K W (editors). Vol. 1, 598pp. (London: Graham and Trotman.)

KNOX, R B O'B. 1984. Lithostratigraphy and depositional history of the late Toarcian sequence at Ravenscar, Yorkshire. *Proceedings of the Yorkshire Geological Society*, Vol. 45, 99–108.

KRINSLEY, D H, and SMITH, D B. 1981. A selective SEM study of grains from the Permian Yellow Sands of north-east England. *Proceedings of the Geologists' Association*, Vol. 92, 189–196.

LEE, A G. 1988. Carboniferous basin configuration of central and northern England modelled using gravity data. 69–84 in *Sedimentation in a synorogenic basin complex: the Upper Carboniferous of northwest Europe*. BESLEY, B M, and KELLING, G (editors). 276pp. (Blackie: Glasgow and London.)

LEEDER, M R. 1982a. *Sedimentology*. 344pp. (London: Allen & Unwin.)

— 1982b. Upper Palaeozoic basins of the British Isles —Caledonide inheritance versus Hercynian plate margin processes. *Journal of the Geological Society of London*, Vol. 139, 479–491.

— 1988. Recent developments in Carboniferous geology: a critical review with implications for the British Isles and NW Europe. *Proceedings of the Geologists' Association*, Vol. 99, 73–100.

LEGRAND-BLAIN, M. 1985. A new genus of Carboniferous spiriferid brachiopod from Scotland. *Palaeontology*, Vol. 28, 567–575.

LEWIS, H C. 1984. *The glacial geology of Great Britain and Ireland*. 469pp. (London: Longmans, Green & Co.)

LORD, A. 1974. Ostracods from the Domerian and Toarcian of England. *Palaeontology*, Vol. 17, 599–622.

MARR, J E. 1921. On the rigidity of northwest Yorkshire. *Naturalist, Hull*, 1921, 63–72.

MARSHALL, W. 1826. Notice of carbonate of copper occurring in the Magnesian Limestone of Newton Kyme, near Tadcaster. *Transactions of the Geological Society of London*, Series 2, Vol. 2, 140–141.

MATTHEWS, B. 1970. Age and origin of aeolian sand in the Vale of York. *Nature, London*, Vol. 227, 1234–1236.

MITCHELL, G H, STEPHENS, J V, BROMEHEAD, C E N, and WRAY, D A. 1947. Geology of the country around Barnsley. *Memoir of the Geological Survey of Great Britain*. Sheet 87 (England and Wales.) 182pp.

MOORE, E W J. 1950. The genus *Sudeticeras* and its distribution in Lancashire and Yorkshire. *Journal of the Manchester Geological Association*, Vol. 2, Pt. 1, 31–50.

MORIGI, A N, and JAMES, J W C. 1984. The sand and gravel resources of the country north-east of Ripon, North Yorkshire: description of 1:25 000 Sheet SE 37 and part of SE 47. *Mineral Assessment Report, British Geological Survey*, No. 143.

MOSS, A. 1985. Coal Board strikes it rich for 21st century. *Yorkshire Evening Post*, 2nd May, 1985.

MOSSOP, G D, and SHEARMAN, D J. 1973. Origins of secondary gypsum rocks. *Transactions of the Institute of Mining and Metallurgy, (Section B: Applied Earth Sciences)*, Vol. 82, B147–B154.

MURRAY, J. 1817. Sulphate of strontia in the banks of the Nidd near Knaresborough. *Transactions of the Geological Society*, Vol. 4, 445.

MURRAY, R C. 1964. Origin and diagenesis of gypsum and anhydrite. *Journal of Sedimentary Petrology*, Vol. 34. 512–523.

NORRIS, A, BARTLEY, D D, and GAUNT, G D. 1971. An account of the deposit of shell marl at Burton Salmon, West Yorkshire. *Naturalist, Hull*, No. 917, 57–63.

O'CONNOR, J. 1964. The geology of the area around Malham Tarn, Yorkshire. *Field Studies*, Vol. 1, 53–82.

PALMER, J, and RADLEY, J. 1961. Gritstone tors of the English Pennines. *Zeitschrift für Geomorphologie*, Band 5, Heft 1, 37–52.

PATTISON, J, SMITH, D B, and WARRINGTON, G. 1973. A review of late Permian and early Triassic biostratigraphy in the British Isles. 220–260 in The Permian and Triassic systems and their mutual boundary. LOGAN, A V, and MILLS, L V (editors). *Memoir of the Canadian Society of Petrological Geology*, No. 2.

— 1978. Upper Permian palaeontology of the Aiskew Bank Farm Borehole, north Yorkshire. *Report of the Institute of Geological Sciences*, No. 78/14. 6pp.

PHAROAH, T C, MERRIMAN, R J, WEBB, P C, and BECKINSALE, R D. 1987. The concealed Caledonides of eastern England: preliminary results of a multidisciplinary study. *Proceedings of the Yorkshire Geological Society*, Vol. 46, 355–369.

PHILLIPS, J. 1865. Note on the geology of Harrogate. *Quarterly Journal of the Geological Society of London*, Vol. 21, 232–235

PLATNAUER, H M. 1888. Note on some crystals of celestine. *Annual Report of the Yorkshire Philosophical Society for 1887*, 34.

POWELL, J H. 1984. Lithostratigraphical nomenclature of the Lias Group in the Yorkshire Basin. *Proceedings of the Yorkshire Geological Society*, Vol. 45, 51–57.

— COOPER, A H, and BENFIELD, A C. 1992. Geology of the country around Thirsk. *Memoir of the British Geological Survey*, Sheet 52 (England and Wales).

RAISTRICK, A. 1932. The correlation of glacial retreat stages across the Pennines. *Proceedings of the Yorkshire Geological Society*, Vol. 22, 199–214.

RAYNER, D H. 1953. The Lower Carboniferous rocks in the north of England. *Proceedings of the Yorkshire Geological Society*, Vol. 28, 231–315.

— and HEMINGWAY, J E. 1974. *The geology and mineral resources of Yorkshire*. 405pp. (Leeds: Yorkshire Geological Society.)

RAMSBOTTOM, W H C. 1966. A pictorial diagram of the Namurian rocks of the Pennines. *Transactions of the Leeds Geological Association*, Vol. 7, 181–184.

— 1974. Dinantian and Namurian. 47–87 in *The geology and mineral resources of Yorkshire*. RAYNER, D H, and HEMINGWAY, J E (editors). 405pp. (Leeds: Yorkshire Geological Society.)

— 1977. Major cycles of transgression and regression (mesothems) in the Namurian. *Proceedings of the Yorkshire Geological Society*, Vol. 41, 261–291.

— CALVER, M A, EAGER, R M C, HODSON, F, HOLLIDAY, D W, STUBBLEFIELD, C J, and WILSON, R B. 1978. A correlation of Silesian rocks in the British Isles. *Special Report of the Geological Society of London*, No. 10.

ROSE, J. 1985. The Dimlington Stadial/Dimlington Chronozone: a proposal for naming the main glacial episode of the Late Devensian in Britain. *Boreas*, Vol. 14, 225–230.

ROWELL, A J, and SCANLON, J E. 1957. The relation between the Yoredale Series and the Millstone Grit on the Askrigg Block. *Proceedings of the Yorkshire Geological Society*, Vol. 31, 79–90.

ROWLEY, D B, RAYMOND, A, PARRISH, J T, LOTTES, A L, SCOTESE, C R, and ZIEGLER, A M. 1985. Carboniferous paleogeographic, phytogenic and paleoclimatic reconstructions. 7–42 in Paleoclimate controls on coal resources in the Pennsylvanian System of North America. PHILLIPS, T L, and CECIL, C B (editors). *International Journal of Coal Geology*, Vol. 5.

SEDGWICK, A. 1829. On the geological relations and internal structure of the Magnesian Limestone, and the lower portions of the New Red Sandstone Series in their range through Nottinghamshire, Derbyshire, Yorkshire and Durham, to the southern extremity of Northumberland. *Transactions of the Geological Society of London*, Series 2, Vol. 3, 37–124.

SMITH, D B. 1968. The Hampole Beds — a significant marker in the Lower Magnesian Limestone of Yorkshire, Derbyshire and Nottinghamshire. *Proceedings of the Yorkshire Geological Society*, Vol. 36, 463–477.

— 1970a. Permian and Trias. 66–91 in The geology of Durham County. JOHNSON, G A L, and HICKLING, G (editors). *Transactions of the Natural History Society of Northumberland*, Vol. 41, No. 1, 152pp.

— 1970b. The palaeogeography of the British Zechstein. 20–23 in *Third symposium on salt*. Vol. 1. RAU, J L, and DELLWIG, L F (editors). 474pp. (Cleveland, Ohio: Northern Ohio Geological Society.)

— 1972. Foundered strata, collapse breccias and subsidence features of the English Zechstein. 255–269 in Geology of saline deposits. *Proceedings of the Hanover Symposium, 1968*. (Earth Sciences, 7) RICHTER-BERNBURG, G (editor). 316pp. (Paris: UNESCO.)

— 1974a. Permian. 115–144 in *The geology and mineral resources of Yorkshire*. RAYNER, D H, and HEMINGWAY, J E (editors). 405pp. (Leeds: Yorkshire Geological Society.)

— 1974b. The stratigraphy and sedimentology of the Permian rocks at outcrop in North Yorkshire. *Journal of Earth Sciences Leeds*, Vol. 8, 365–386.

— 1976. The Upper Permian sabkha sequence at Quarry Moor, Ripon, Yorkshire. *Proceedings of the Yorkshire Geological Society*, Vol. 40, 639–652.

— 1979. Rapid marine transgressions and regressions of the Upper Permian Zechstein Sea. *Journal of the Geological Society of London*, Vol. 136, 155–156.

— 1980. The evolution of the English Zechstein basin. *Contributions to Sedimentology*, Vol. 9, 7–34.

— 1981. Bryozoan–algal patch reefs in the Upper Permian Lower Magnesian Limestone of Yorkshire, northeast England. 187–202 in European fossil reef models. TOOMEY, D F (editor). *Special Publication of the Society of Economic Paleontologists and Mineralogists*, No. 30.

— 1989. The late Permian palaeogeography of north-east England. *Proceedings of the Yorkshire Geological Society*, Vol. 47, 285–312.

— In press. Marine Permian review. *Nature Conservancy Council*.

— BRUNSTROM, R G S, MANNING, P E, SIMPSON, S, and SHOTTON, F W. 1974. A correlation of the Permian rocks of the British Isles. *Special Report of the Geological Society of London*, No. 5. 45pp.

— and FRANCIS, E A. 1967. The geology of the country between Durham and West Hartlepool. *Memoir of the Geological Survey of Great Britain*, Sheet 27 (England and Wales). 354pp.

— HARWOOD, G M, PATTISON, J, and PETTIGREW, T. 1986. A revised nomenclature for Upper Permian strata in eastern England. 9–17 in The English Zechstein and related topics. HARWOOD, G M, and SMITH, D B (editors). *Special Publication of the Geological Society of London*, No. 22. 244pp.

SMITH, E G, and WARRINGTON, G. 1971. The age and relationships of the Triassic rocks assigned to the lower part of the Keuper in north Nottinghamshire, north-west Lincolnshire and south Yorkshire. *Proceedings of the Yorkshire Geological Society*, Vol. 38, 201–227.

— RHYS, G H, and GOOSENS, R F. 1973. Geology of the country around East Retford, Worksop and Gainsborough. *Memoir of the Geological Survey of Great Britain*, Sheet 101 (England and Wales). 348pp.

SMITH, W. 1821. Geological map of Yorkshire, scale approximately two and three quarters miles per inch.

SMITHELLS, A. 1919. *Report of a scientific survey of Harrogate Spa waters*. 114pp. (Leeds: Chorley and Pickersgill.)

SMITHSON, F. 1931. The Triassic sandstones of Yorkshire and Durham. *Proceedings of the Geologists' Association*, Vol. 42, 125–156.

SOPER, N J, WEBB, B C, and WOODCOCK, N H. 1987. Late Caledonian (Acadian) transpression in north-west England: timing, geometry and geotectonic significance. *Proceedings of the Yorkshire Geological Society*, Vol. 46, 175–192.

STANCZYSZYN, R. 1982. The sand and gravel resources of the country around Tholthorpe, North Yorkshire: description of 1:25 000 Sheet SE 46. *Mineral Assessment Report, Institute of Geological Sciences*, No. 88.

STEELE, R P. 1983. Longitudinal drag in the Permian Yellow Sands of northeast England. 543–550 in Eolian sediments and processes. BROOKFIELD, M E, and AHLBRANDT, T S (editors). *Developments in Sedimentology*, No. 38. 660pp.

STEPHENS, J V, MITCHELL, G H, and EDWARDS, W. 1953. Geology of the country between Bradford and Skipton. *Memoir of the Geological Survey of Great Britain*, Sheet 69 (England and Wales). 180pp.

STEWART, F H. 1963. The Permian Lower Evaporites of Fordon, Yorkshire. *Proceedings of the Yorkshire Geological Society*, Vol. 34, 1–44.

STURM, M, and MATTER, A. 1978. Turbidites and varves in Lake Brienz (Switzerland); deposition of clastic detritus by density currents. 147–168 in Modern and ancient lake sediments. MATTER, A, and TUCKER, M E (editors). *Special Publication of the International Association of Sedimentologists*, No. 2.

SUGDEN, D E, and JOHN, B S. 1976. *Glaciers and landscape*. 375pp. (London: Arnold.)

TATE, R, and BLAKE, J F. 1876. *The Yorkshire Lias*. 476pp. (London: John van Voorst.)

TAYLOR, B J, BURGESS, I C, LAND, D H, MILLS, D A C, SMITH, D B, and WARREN, P T. 1971. *British regional geology: Northern*

*England* (4th edition). 121pp. (London: HMSO for Institute of Geological Sciences.)

TAYLOR, J C M, and COLTER, V S. 1975. Zechstein of the English Sector of the southern North Sea basin. 249–263 in *Petroleum and the continental shelf of north-west Europe*, Vol. 1 Geology. WOODLAND, A W (editor). 501pp. (Barking: England: Applied Science Publishers Ltd.)

TILLOTSON, E. 1932. The glacial geology of Nidderdale. *Proceedings of the Yorkshire Geological Society*, Vol. 22, 215–228.

TURNER, P, and MAGARITZ, M. 1986. Chemical and isotopic studies of a core of Marl Slate from NE England: influence of freshwater influx into the Zechstein Sea. 19–29 *in* The English Zechstein and related topics. HARWOOD, G M, and SMITH, D B (editors). *Special Publication of the Geological Society of London*, No. 22.

TUTE, Rev. J S. 1868. On certain natural pits in the neighbourhood of Ripon. *Geological Magazine*, Vol. 5, 178–179.

— 1870. On certain natural pits in the neighbourhood of Ripon. *Proceedings of the Yorkshire Geological and Polytechnic Society*, Vol. 5, 2–7.

— 1886. The Cayton Gill Beds. *Proceedings of the Yorkshire Geological and Polytechnic Society*, Vol. 9, 265–267

— 1890. Notes on some singular cavities in the Magnesian Limestone. *Proceedings of the Yorkshire Geological and Polytechnic Society*, Vol. 11, 182–184.

— 1892. On a Permian conglomerate bed at Markington. *Proceedings of the Yorkshire Geological and Polytechnic Society*, Vol. 12, 72–73.

— 1894. On some singular cavities in the Magnesian Limestone. *Proceedings of the Yorkshire Geological and Polytechnic Society*, Vol. 12, 245–246.

VERSEY, H C. 1925. The beds underlying the Magnesian Limestone in Yorkshire. *Proceedings of the Yorkshire Geological Society*, Vol. 20, 200–214.

WARRINGTON, G. 1974. Triassic. 145–160 in *The geology and mineral resources of Yorkshire*. RAYNER, D H, and HEMINGWAY, J E (editors). 405pp. (Leeds: Yorkshire Geological Society.)

— AUDLEY-CHARLES, M G, ELLIOTT, R E, EVANS, W B, IVIMEY-COOK, H C, KENT, P, ROBINSON, P L, SHOTTON, F W, and TAYLOR, F M. 1980. A correlation of Triassic rocks in the British Isles. *Special Report of the Geological Society of London*, No. 13, 78pp.

WHITTAKER, A. 1985. *Atlas of onshore sedimentary basins in England and Wales. Post-Carboniferous tectonics and stratigraphy. Special Publication of the British Geological Survey.* (Glasgow and London: Blackie.)

— HOLLIDAY, D W, and PENN, I E. 1985. Geophysical logs in British stratigraphy. *Special Report of the Geological Society of London*, No. 18.

WILSON, A A. 1960. The Carboniferous rocks of Coverdale and adjacent valleys in the Yorkshire Pennines. *Proceedings of the Yorkshire Geological Society*, Vol. 32, 285–316.

— 1977. The Namurian rocks of the Fewston area. *Transactions of the Leeds Geological Association*, Vol. 9, 1–42.

— and THOMSON, A T. 1965. The Carboniferous succession in the Kirkby Malzeard area, Yorkshire. *Proceedings of the Yorkshire Geological Society*, Vol. 35, 203–227, pl. 15.

ZIEGLER, P A. 1982. *Geological atlas of Western and Central Europe.* 130pp. 40 loose-bound enclosures. (Shell International Petroleum Maatschappij, B V, distributed by Elsevier.)

# APPENDIX 1

## Abstracts of selected borehole logs

The boreholes are identified by their registered numbers in the BGS 1:10 000 sheet registration system.

### Ellenthorpe No. 1 Borehole (SE 46 NW /7) SE 4233 6703

Surface level +c.15 m above OD, depths from rotary table level of +18.3 m above OD. Drilled in 1945–1946.

The Permian and Triassic logs are based on the original oil company records and a re-examination of the chipping samples for most of the Permian.

| | *Thickness* | *Depth* |
| | m | m |
| **Pleistocene and recent** | | |
| UNDIVIDED | 38 | 38 |
| **Triassic** | | |
| SHERWOOD SANDSTONE GROUP | 125 | 163 |
| **Permian** | | |
| UPPER MARL | 37 | 200 |
| UPPER MAGNESIAN LIMESTONE | 14 | 214 |
| MIDDLE MARL | 40 | 254 |
| LOWER MAGNESIAN LIMESTONE | 61 | 315 |
| BASAL PERMIAN SANDS | 2 | 317 |

Summary of Lower Carboniferous (Dinantian) strata based on a re-examination of chipping and core samples.

### Carboniferous

### Dinantian

| | | |
|---|---|---|
| Limestone; pale grey, dolomitic packstone grainstone, with sporadic partings of dark grey, silty, calcareous mudstone; the presence of the foraminifera *Brunsia* sp. and *Paraarchaediscus* sp. at 318.5 and 335.3 m show this interval to be of Carboniferous age (probably Asbian), not Permian as suggested on the original completion log | 63 | 380 |

Limestone; medium to dark grey packstone/ wackestone, silty and sandy in part, dolomitised in places, with interbedded grey, calcareous, fossiliferous siltstone and sandstone. This interval yielded *Archaediscus* sp. stage *angulatus* at 385.09 m, together with the dasyclad *Koninckopora*. These allow correlation with the Cf6a-γ foraminiferal zones and indicate a probable Asbian age. However, in limestone turbidites, *Koninckopora* may be reworked into younger strata and may lead to a misinterpretation of the age. At the interval below 427.91 m, the sequence yielded

| | | |
|---|---|---|
| *Pojarkovella*, a guide fossil to the Cf5–Cf6β zonal interval; the lack of definite Cf6 zone guide fossils suggests a Cf5 zone assignation, which is probably no older than of Holkerian age. *Paraarchaediscus* sp. at the *concavus* stage of coiling was also present. | 110 | 490 |
| Sandstone, pale grey to buff, fine- to medium-grained, micaceous, carbonaceous ripple-cross-bedded, with interbedded, grey, micaceous siltstone; sporadic thin limestone bands | 150 | 640 |
| Siltstone, dark grey, micaceous, with sporadic thin beds of limestone and calcareous sandstone | 120 | 760 |
| Siltstone, dark grey, micaceous, calcareous in part, with interbedded, fine-grained, carbonaceous sandstone and thin beds of crinoidal limestone and calcareous sandstone; at 789.08 m this interval yielded *Paraarchaediscus* sp. at the *concavus* stage of coiling, which in the context of the overlying and underlying intervals suggests a Holkerian age. Below 791.83 down to 1095.7 m the sequence contains abundant primitive archaediscids, notably *Glomodiscus*, cf. *Uralodiscus* and *Paraarchaediscus* sp. at the *involutus* stage of coiling, indicating a Cf4b–d correlation and an Arundian age. | 80 | 840 |
| Siltstone, dark grey, micaceous, calcareous in part, with interbedded thin limestone bands and calcareous mudstones; fossils as above indicating an Arundian age. | 175 | 1015 |
| Limestone, medium to dark grey; mainly crinoidal/algal grainstones, with interbedded packstone and wackestone; mudstone partings in upper half; fossils as above indicating an Arundian age. | 81.7 | 1096.7 |
| Bottom of borehole | | 1096.7 |

The top of the Dinantian sequence is taken at a higher level in the borehole than it was by Falcon and Kent (1960). Stages are based on thin-section identification of foraminifera by Drs N J Riley and A R E Strank. Thicknesses given are uncorrected for dip.

## Burton Leonard Borehole (SE 36 SW/16) SE 3279 6462

Drilled in 1959 Surface level +c.68.0 m above OD. Record based on the drillers log

| | Thickness m | Depth m |
|---|---|---|
| **Glacial deposits** | | |
| Stony soil | 0.46 | 0.46 |
| **Permian** | | |
| UPPER MARL | | |
| Red marl | 1.98 | 2.44 |
| UPPER MAGNESIAN LIMESTONE | | |
| Limestone | 8.84 | 11.28 |
| MIDDLE MARL | | |
| Red marl | 5.48 | 16.76 |
| Limestone (Kirkham Abbey Formation?) | 1.83 | 18.59 |
| Soft red marl | 3.97 | 22.56 |
| Hard red marl and gypsum | 9.44 | 32.00 |
| Soft limestone | 0.92 | 32.92 |
| Red marl | 4.88 | 37.80 |
| Soft red and blue marl | 1.21 | 39.01 |
| LOWER MAGNESIAN LIMESTONE? | | |
| Limestone | 3.67 | 42.68 |
| Bottom of borehole | | 42.68 |

## Farnham IGS Borehole (SE 35 NE/27) SE 3469 5996

Drilled in March 1979
Surface level +42 m OD

Details of the borehole are shown in Figure 8, and further information is given by Burgess and Cooper (1980b).

## Crimple Beck Borehole (SE 25SE/12) SE2728 5186

Drilled in 1982
Surface level +132 m above OD

Coring was continuous from 5 m. The average dip of the strata was about 25°, (increasing locally to 60°); thicknesses given are uncorrected for dip.

| | Thickness in borehole m | Depth m |
|---|---|---|
| **Pleistocene and recent** | | |
| Clay, yellow-brown, with sandstone cobbles (not sampled) (Head and Till) | 3.00 | 3.00 |
| **Carboniferous** | | |
| **Namurian (Pendleian Stage)** | | |
| Mudstone, dark grey, with laminae and thin graded beds of crinoidal limestone towards base; sporadic marine fossils, including *Tumulites pseudobilinguis* | 35.80 | 38.80 |
| HARROGATE ROADSTONE: Limestone, pale grey, fine- to coarse-grained, crinoidal, silicified, in beds 0.05 to 1.00 m thick; sandy at base; some beds graded, some laminated or with low-angle cross-stratification. | 20.60 | 59.40 |
| **Dinantian (Brigantian Stage)** | | |
| Mudstone, dark grey, with laminae and thin graded beds of crinoidal limestone; 1.6 m | | |

| | | |
|---|---|---|
| debris bed at base; sporadic marine fossils, including *Sudeticeras* sp. | 11.40 | 70.80 |
| Mudstone, grey, with laminae and bands of fine-grained sandstone. | 3.95 | 74.75 |
| Sandstone, grey, fine- to medium-grained. | 2.85 | 77.60 |
| Mudstone, dark grey, with laminae and bands of fine-grained sandstone | 9.95 | 87.55 |
| HARLOW HILL SANDSTONE: Sandstone, pale grey, fine- to coarse-grained, in beds 0.05 to 5.3 m thick; some beds massive with mudstone clasts, some graded, some laminated or low-angle cross-stratified; interbedded with micaceous siltstones and fine-grained sandstones. | 46.55 | 134.10 |
| Mudstone, dark grey, with laminae and bands of fine-grained sandstone. | 4.20 | 138.30 |
| Mudstone, dark grey; debris beds with mudstone clasts and shelly fossils. | 5.75 | 144.05 |
| Mudstone, dark grey, with laminae and bands of fine-grained sandstone. | 29.45 | 173.50 |
| Fault breccia | 0.30 | 173.80 |
| Sandstone, pale grey, thinly bedded, fine-grained; partly flaser-bedded, with ripples and mud drapes; partly laminated; sporadic thicker graded beds. | 14.70 | 188.50 |
| Bottom of borehole | | 188.50 |

## Spellow Hill Borehole (SE 36 SE/10) SE 3816 6234

Drilled in 1941
Surface level c.+54 m OD

Lithological descriptions taken from the drillers log

| | Thickness m | Depth m |
|---|---|---|
| **Pleistocene and Permian undivided** | | |
| Red clay | 3.35 | 3.35 |
| **Permian** | | |
| UPPER MAGNESIAN LIMESTONE | | |
| Limestone (with water) | 9.60 | 12.95 |
| MIDDLE MARL | | |
| Blue Marl | 3.05 | 16.00 |
| Red Marl | 5.64 | 21.64 |
| Marl | 0.15 | 21.79 |
| Limestone (Kirkham Abbey Formation or gypsum?) | 5.34 | 27.13 |
| Blue Marl | 2.74 | 29.87 |
| Grey shale with seams of gypsum | 1.83 | 31.70 |
| Grey shale | 3.28 | 34.98 |
| 'Limestone' Marl, stronger near bottom (Lower Magnesian Limestone or gypsum in Middle Marl) | 18.36 | 53.34 |
| LOWER MAGNESIAN LIMESTONE | | |
| Yellow Limestone | 12.80 | 66.14 |
| Limestone | 14.33 | 80.47 |
| Hard limestone and pyrites | 1.83 | 82.30 |
| Hard limestone with pyrites and galena | 0.61 | 82.91 |
| Hard honeycombed limestone | 9.75 | 92.66 |
| Hard limestone broken in thin beds | 10.67 | 103.33 |
| Dark limestone | 2.44 | 105.77 |
| LOWER MARL | | |
| Blue shale | 13.41 | 119.18 |

BASAL BRECCIA
| | | |
|---|---|---|
| Conglomerate | 2.44 | 121.62 |

**Carboniferous**

| | | |
|---|---|---|
| Red shale with ironstone nodules | 5.79 | 127.41 |
| Grey sandstone | 4.87 | 132.28 |
| Grey shale | 2.29 | 134.57 |
| Coal | 0.51 | 135.08 |
| Grey shale | 1.52 | 136.60 |
| Coal, bright and very light | 0.33 | 136.93 |
| Blackstone and coal | 0.06 | 136.99 |
| Bright coal | 0.13 | 137.12 |
| Blackstone and coal | 0.08 | 137.20 |
| Grey shale | 0.26 | 137.46 |
| Hard grey sandstone | 0.92 | 138.38 |
| Soft seatearth clay with carbonaceous rootlets | 1.22 | 139.60 |
| Grey shale | 1.83 | 141.43 |
| Hard grey sandstone | 2.44 | 143.87 |
| Grey shale with ironstone nodules | 3.65 | 147.52 |
| Grey shale with siltstone partings | 5.95 | 153.47 |
| Dark grey shale with sandstone beds | 2.13 | 155.60 |
| Coal, coarse | 0.88 | 156.48 |
| Hard grey sandstone | 1.22 | 157.70 |
| Bottom of borehole | | 157.70 |

**Stockeld Borehole** (SE 34 NE/16) SE 3803 4945
Drilled in 1951
Surface level c.40.46 m above OD

Log based on a re-examination of the cores by Dr N J Riley

| | *Thickness* | *Depth* |
|---|---|---|
| | m | m |

**Carboniferous**

| | | |
|---|---|---|
| UPPER PLOMPTON GRIT: Sandstone, coarse-grained and micaceous (grit) | 8.53 | 8.53 |
| Mudstone, grey, silty, with sandstone laminae and very thin beds, and ironstone nodules | 24.08 | 32.61 |
| ECCUP MARINE BAND: (Reticuloceras coreticulatum Marine Band, $R1_{c4}$) Mudstone, grey, soft and pyritous, with *Crurithyris* sp., *Lingula mytilloides*, *Orbiculoidea* sp., *Anthraconielio* sp., *Productus carbonarius*, *Rugusochonetes* sp., *Euphemites* sp., *Euchondria?*, *Schizodus* sp., *Coleolus namurcensis*. | 4.58 | 37.19 |

| | | |
|---|---|---|
| LOWER PLOMPTON GRIT: Sandstone, fine-grained and massive | 1.07 | 38.25 |
| Mudstone, sandy | 1.07 | 39.32 |
| Mudstone, micaceous, with sandstone beds | 2.44 | 41.76 |
| Sandstone, coarse-grained | 18.29 | 60.05 |
| Mudstone and sandy siltstone with subordinate sandstone beds | 2.43 | 62.48 |
| Sandstone, medium- to coarse-grained | 1.83 | 64.31 |
| Siltstone and mudstone, interlaminated | 1.83 | 66.14 |
| Sandstone, medium-grained | 3.20 | 69.34 |
| Mudstone, grey, with beds of sandy siltstone and subordinate beds of sandstone; fault? at 94.79 m | 28.20 | 97.54 |
| MARINE BAND: (Reticuloceras reticulatum Marine Band, $R1c_1$?) Mudstone, grey with beds of calcareous mudstone and impure limestone, fossiliferous, with *Lingula mytilloides*, *Productus carbonarius*, *Spirifer bisulcatus?*, *Aviculopecten*, *Parametacoceras pulcher*, *Reticuloceras reticulatum* and *Vallites striolatus*. | 4.57 | 102.11 |
| ADDLETHORPE GRIT: Sandstone, grey, fine- to coarse-grained | 6.70 | 108.81 |
| Mudstone and siltstone with subordinate sandstone beds | 21.03 | 129.84 |
| MARINE BAND: Mudstone and siltstone pyritous, with ironstone nodules, and marine fossils including: *Orbiculoidea nitida*, *Caneyella* sp., productoid and other shell fragments, and *Vallites* sp.? | 27.44 | 157.28 |
| Mudstone and siltstone with subordinate sandstone beds | 12.19 | 169.47 |
| CAYTON GILL SHELL BED? Sandstone, hard, white and massive, with shaly wisps | 4.27 | 173.74 |
| Mudstone with scattered pyrite and subordinate beds of sandstone | 9.14 | 182.88 |
| Bottom of borehole | | 182.88 |

# APPENDIX 2

## Geological Survey photographs

Copies of these photographs may be seen in the library of the British Geological Survey, Keyworth, Nottinghan, NG12 5GG. Prints and slides may be purchased at a fixed tariff. 'A' Series photographs monochrome 1930–1938; 'L' Series photographs colour 1975-1980. Grid references of views refer to the viewpoint.

### CARBONIFEROUS

L 1387   The Harrogate Roadstone; well-bedded, partially silicified, crinoidal, bioclastic limestones [2808 5236]

L 1388   The Harrogate Roadstone; well-bedded, partially silicified, crinoidal, bioclastic limestones [2808 5236]

L 1389   Fault in Harlow Hill Sandstone [2855 5374]

L 1390   Mudstones with interbedded thin sandstones overlying the Harlow Hill Sandstone [2855 5374]

L 2916   Flat-lying, cross-bedded, pebbly sandstone. (Lower Plompton Grit) Fountains Abbey, Near Ripon [2764 6836]

L 2917   Cross-bedded, yellow stained, fine- to coarse-grained, pebbly sandstone. (Lower Plompton Grit) Fountains Abbey, Near Ripon [2764 6838]

L 1822   Lover's Leap, crags of Upper Plompton Grit, Plompton Rocks, near Knaresborough [3552 5363]

L 1823   Lover's Leap, crags of Upper Plompton Grit, Plompton Rocks, near Knaresborough [3552 5363]

L 1824   Crags of Upper Plompton Grit, Plompton Rocks, near Knaresborough [3547 5379]

L 1825   Crags of Upper Plompton Grit, Plompton Rocks, near Knaresborough [3547 5397]

L 1826   The escarpment of the Upper Plompton Grit at Plompton Park, near Knaresborough [3549 5393]

L 1827   Naturally sculptured crags and stacks of Upper Plompton Grit, Crimple Beck [356 527]

L 1828   Naturally sculptured crags and stacks of Upper Plompton Grit, Crimple Beck [356 527]

L 1829   Naturally sculptured crags and stacks of Upper Plompton Grit, Crimple Beck [365 527]

### PERMIAN

A 5225   Unconformable junction of Lower Magnesian Limestone on Upper Plompton Grit, Newsome Bridge Quarry, North Deighton [3789 5143]

A 5226   Unconformable junction of Lower Magnesian Limestone on Upper Plompton Grit, Newsome Bridge Quarry, North Deighton [3789 5143]

A 5227   Unconformable junction of Lower Magnesian Limestone on Upper Plompton Grit, Newsome Bridge Quarry, North Deighton [3789 5143]

A 5228   Unconformable junction of Lower Magnesian Limestone on Upper Plompton Grit, Newsome Bridge Quarry, North Deighton [3789 5143]

A 5229   Thin-bedded limestone of the Lower Magnesian Limestone, Plompton Quarry, near Grimbald Bridge [365 558]

A 5230   Thin-bedded limestone of the Lower Magnesian Limestone, Plompton Quarry, near Grimbald Bridge [3650 5585]

A 5231   Cliff of Lower Magnesian Limestone, Knaresborough [c. 3515 5637]

A 5232   Lower Magnesian Limestone, Knaresborough [c. 3515 5637]

A 5233   Restored doorway of Goldsborough Church showing weathering of Magnesian Limestone Church [3845 5610]

A 5234   Restored doorway of Goldsborough Church showing weathering of Magnesian Limestone Church [3845 6510]

A 5235   Lower Magnesian Limestone, Old Quarry, Burton Leonard [c. 3235 6298]

A 5236   Lower Magnesian Limestone, Old Quarry, Burton Leonard [c. 3235 6298]

A 7547   Lower Magnesian Limestone resting unconformably on Upper Plompton Grit, Newsome Bridge Quarry, North Deighton [3789 5143]

A 7548   Lower Magnesian Limestone resting unconformably on Upper Plompton Grit, Newsome Bridge Quarry, North Deighton [3789 5143]

L 1801   Upper Magnesian Limestone in disused quarry near Burton Leonard [3319 6345]

L 1802   Landslip in the Middle Marl near Goldsborough Mill, Knaresborough [3699 5608]

L 1803   The Upper Permian sabkha sequence in the Lower Magnesian Limestone at Quarry Moor, Ripon [3080 6921]

L 1804   The Upper Permian sabkha sequence in the Lower Magnesian Limestone at Quarry Moor, Ripon [3080 6921]

L 1805   The Upper Permian sabkha sequence in the Lower Magnesian Limestone at Quarry Moor, Ripon [3080 6921]

L 1806   Lower Magnesian Limestone, Bunkers Hill Quarry, Knaresborough [3513 5660]

L 1807   Lower Magnesian Limestone, Bunkers Hill Quarry, Knaresborough [3513 5660]

L 1808   The House in the Rock, Lower Magnesian Limestone, Knaresborough [3513 5645]

L 1809   Chapel of Our Lady of the Crag, Lower Magnesian Limestone, Knaresborough [3513 5645]

L 1810   Chapel of Our Lady of the Crag, Lower Magnesian Limestone, Knaresborough [3513 5645]

L 1811   Lower Magnesian Limestone in Knaresborough gorge [3515 5642]

L 1812   Saint Robert's Cave, Lower Magnesian Limestone, Knaresborough [3610 5609]

L 1813   Saint Robert's Cave, Lower Magnesian Limestone, Knaresborough [3610 5609]

L 1814   Unconformity between the Lower Magnesian Limestone and the Upper Plompton Grit at Abbey Crags, Knaresborough [3550 5583]

L 1815   Unconformity between the Lower Magnesian Limestone and the Upper Plompton Grit near the Abbey, Knaresborough [3575 5572]

L 1816   Unconformity between the Lower Magnesian Limestone and the Upper Plompton Grit near the Abbey, Knaresborough [3575 5572]

L 1817   Unconformity between the Lower Magnesian Limestone and Upper Plompton Grit at Grimbald Crag, Knaresborough [3605 5576]

L 1818   Unconformity between the Lower Magnesian Limestone and the Upper Plompton Grit at Newsome Bridge Quarry, North Deighton [3788 5146]

L 1819   Breccia at the base of the Lower Magnesian Limestone, resting unconformably on the Upper Plompton Grit at Newsome Bridge Quarry, North Deighton [3788 5146]

L 1820   Breccia at the base of the Lower Magnesian Limestone, resting unconformably on the Upper Plompton Grit at Newsome BridgeQuarry, North Deighton [3788 5146]

L 1821   Breccia at the base of the Lower Magnesian Limestone, resting unconformably on the Upper Plompton Grit at Newsome Bridge Quarry, North Deighton [3788 5146]

L 3078   Lower Magnesian Limestone exposed in the steep-sided valley of the River Skell, which was cut by glacial meltwater [2838 6914]

## JURASSIC

L 3088   Fault-bounded Lower Jurassic escarpment, Mill Lane, Easingwold [5385 6961]

## QUATERNARY

L 1795   Distorted glacial deposits at Grafton gravel quarry [4207 6295]

L 1796   Gravel on contorted sand and clay at Grafton gravel quarry [4214 6290]

L 1797   Till overlying stratified sand deposits at Grafton gravel quarry [4204 6302]

L 1798   Sandy Till on gravelly boulder clay at Grafton gravel quarry [4200 6310]

L 1799   Gravelly Till at Grafton gravel quarry [4204 6295]

L 1800   Inferred ice-contact slope near Grafton [4112 6338]

L 2427   Drumlin-shaped ridges in the village of Coneythorpe [3911 5919]

L 2428   Drumlin-shaped ridges in the village of Coneythorpe [3911 5919]

L 2429   Morainic ridge between Flaxby and Allerton Park [4074 5723]

L 2930   Rounded hills of sand (glacial deposits; Devensian) north of Ferrensby [3669 6092]

L 2431   Rounded hills of sand (glacial deposits; Devensian) north of Ferrensby [3670 6075]

L 2432   Glacial drainage channel cut into the Upper Magnesian Limestone near Copgrove [346 630]

L 2433   Glacial drainage channel cut into the Upper Magnesian Limestone near Copgrove [346 630]

L 2434   Glacial drainage channel cut into the Lower Magnesian Limestone near Farnham [3516 6158]

L 2435   Glacial drainage channel cut into the Lower Magnesian Limestone near Farnham [3516 6158]

L 2915   Ten Mile Hill, west of Tholthorpe, a broad esker of glacial deposits protruding from flat glacial lake deposits [4556 6694]

L 2918   Remains of chapel of St Michael de Monte on top of How Hill moraine [2767 6668]

L 2919   Glacial drainage channel cut into Lower Magnesian Limestone, River Skell, Studley Royal, Near Ripon [2870 6916]

L 3035   Laminated clay (Glacial Lake Deposits) at Littlethorpe Pottery, near Ripon [3264 6811]

L 3036   Laminated clay (Glacial Lake Deposits) at Littlethorpe Pottery, near Ripon [3264 6811]

L 3037   Mr Curtis of Littlethorpe Pottery, throwing a pot [3248 6813]

L 3038   Mr Curtis of Littlethorpe Pottery, finishing a pot [3248 6813]

L 3079   Section in old sand and gravel pit between Tholthorpe and Flawith showing Glacial Sand and Gravel; part of an esker complex [c. 4760 6620]

L 3080   View from Myton Moor to esker complex of Ten Mile Hill [4556 6703]

L 3081   Glacial Sand and Gravel in a pit at Rice Lane, near Aldwark [4606 6474]

L 3082   Glacial Sand and Gravel in a pit at Rice Lane, near Aldwark [4606 6474]

L 3083   Glacial Sand and Gravel in a pit at Rice Lane, near Aldwark [4606 6474]

L 3084   Laminated clay (Glacial Lake Deposits) at Alne clay pit, near Easingwold [5193 6630]

# FOSSIL INDEX

# GENERAL INDEX

Page numbers in italics refer to figures.
Page numbers in bold refer to tables.
Page numbers followed by 'P' refer to
plates.

**BRITISH GEOLOGICAL SURVEY**

Keyworth, Nottingham NG12 5GG
(0602) 363100

Murchison House, West Mains Road, Edinburgh
EH9 3LA    031-667 1000

London Information Office, Natural History Museum
Earth Galleries, Exhibition Road, London SW7 2DE
071-589 4090

The full range of Survey publications is available through the Sales Desks at Keyworth and at Murchison House, Edinburgh, and in the BGS London Information Office in the Natural History Museum Earth Galleries. The adjacent bookshop stocks the more popular books for sale over the counter. Most BGS books and reports are listed in HMSO's Sectional List 45, and can be bought from HMSO and through HMSO agents and retailers. Maps are listed in the BGS Map Catalogue, and can be bought BGS approved stockists and agents as well as direct from BGS.

*The British Geological Survey carries out the geological survey of Great Britain and Northern Ireland (the latter as an agency service for the government of Northern Ireland), and of the surrounding continental shelf, as well as its basic research projects. It also undertakes programmes of British technical aid in geology in developing countries as arranged by the Overseas Development Administration.*

*The British Geological Survey is a component body of the Natural Environment Research Council.*

HMSO publications are available from:

**HMSO Publications Centre**
(Mail, fax and telephone orders only)
PO Box 276, London SW8 5DT
Telephone orders 071-873 9090
General enquiries 071-873 0011
Queueing system in operation for both numbers
Fax orders 071-873 8200

**HMSO Bookshops**
49 High Holborn, London WC1V 6HB
(counter service only)
071-873 0011    Fax 071-873 8200
258 Broad Street, Birmingham B1 2HE
021-643 3740    Fax 021-643 6510
33 Wine Street, Bristol BS1 2BQ
0272-264306    Fax 0272-294515
9 Princess Street, Manchester M60 8AS
061-834 7201    Fax 061-833 0634
16 Arthur Street, Belfast BT1 4GD
0232-238451    Fax 0232-235401
71 Lothian Road, Edinburgh EH3 9AZ
031-228 4181    Fax 031-229 2734

**HMSO's Accredited Agents**
(see Yellow Pages)

*And through good booksellers*